TEUBNER-TEXTE zur Physik Band 34

S. Moss de Oliveira / P. M. C. de Oliveira /
D. Stauffer

Evolution, Money, War, and Computers

TEUBNER-TEXTE zur Physik

This regular series includes the presentation of recent research developments of strong interest as well as comprehensive treatments of important selected topics of physics. One of the aims is to make new results of research available to graduate students and younger scientists, and moreover to all people who like to widen their scope and inform themselves about new developments and trends.

A larger part of physics and applications of physics and also its application in neighbouring sciences such as chemistry, biology and technology is covered. Examples for typical topics are: Statistical physics, physics of condensed matter, interaction of light with matter, mesoscopic physics, physics of surfaces and interfaces, laser physics, nonlinear processes and selforganization, ultrafast dynamics, chemical and biological physics, quantum measuring devices with ultimately high resolution and sensitivity, and finally applications of physics in interdisciplinary fields.

Evolution, Money, War, and Computers

Non-Traditional Applications of Computational Statistical Physics

By Prof. Dr. Suzana Moss de Oliveira
Prof. Dr. Paulo Murilo C. de Oliveira
Universidade Federal Fluminense
Niterói, Rio de Janeiro

Prof. Dr. Dietrich Stauffer
Cologne University

Springer Fachmedien Wiesbaden GmbH

Suzana Moss de Oliveira: Born in 1955 in Rio de Janeiro (Brazil). Tenured professor at Institute of Physics, Universidade Federal Fluminense (Niterói, Rio de Janeiro) since 1985; PhD in 1989 also in Niterói; at present deputy head of the Institute. Introduced to biological ageing by watching Dietrich Stauffer in 1994, frequent visitor to Laboratoire de Physique et Mécanique des Milieux Hétérogènes (ESPCI, Paris) to work on interdisciplinary subjects.

Paulo Murilo C. de Oliveira: Born in 1949 in Rio de Janeiro (Brazil). 100 papers, one previous book. Full professor of Physics, Universidade Federal Fluminense (Niterói, Rio de Janeiro) since 1994, formerly tenured professor at the same place, since 1981; PhD in 1980 from the Catholic University of Rio de Janeiro. Member of the Brazilian Academy of Sciences, former general secretary and member of the deliberative council of the Brazilian Physical Society, and current member of the IUPAP Comission for Computational Physics. Member of the editorial board of Brazilian Journal of Physics, International Journal of Modern Physics C, Revista Boliviana de Física and Theory in Biosciences. Scientific consultant for Brazilian agencies, including CNPq for research programs and CAPES for graduate studies, as well as a dozen international physics journals.

Dietrich Stauffer: Born in 1943 in Bonn (Germany), PhD 1970 in Munich, Habilitation 1975 in Saarbrücken (with K. Binder), since 1977 tenured professor in Cologne; French-German and Canadian-German research prizes 1986, 1993 and 1999. 400 papers, 3 previous books: on percolation, on theoretical physics, on computational physics. Lectures mostly for future high-school teachers; managing editor for Annual Reviews of Computational Physics and International Journal of Modern Physics C.

Gedruckt auf chlorfrei gebleichtem Papier.

Die Deutsche Bibliothek – CIP-Einheitsaufnahme

Oliveira, Suzana Moss de:
Evolution, money, war, and computers : non-traditional applications or computational statistical physics / Suzana Moss de Oliveira ;
Paulo Murilo C. de Oliveira ; Dietrich Stauffer. –
Stuttgart ; Leipzig : Teubner, 1999
(Teubner-Texte zur Physik ; Bd. 34)

ISBN 978-3-519-00279-6 ISBN 978-3-322-91009-7 (eBook)
DOI 10.1007/978-3-322-91009-7

Ursprünglich erschienen bei B. G. Teubnuer Stuttgart . Leipzig 1999.

Umschlaggestaltung: E. Kretschmer, Leipzig

Preface

Physics has always influenced other fields, from the mirrors of Archimedes which were supposed to ignite the sails of enemy warships, to the hydrogen bombs which could destroy São Paulo and Düsseldorf. More recently, (computational) statistical physics has been applied to biology, economics, world politics, etc. This book presents some examples in this direction to explain these methods. We do not claim to present a textbook of biology, or a solution of economic problems; we merely explain how physicists like us did it, in order to encourage others to try similar interdisciplinary work. The theories presented here may not be the best ones for the problems but should be the ones closest to statistical physics methods. In most fields we give only the general idea to guide the reader to the original literature; for the Penna model of biological ageing we try to give a more complete overview, including computer programs, so that readers can start their own research immediately in this field. The selection of material follows our own experience, and even with these selected applications we would appreciate if readers would inform us about overlooked literature. We prove our ignorance, if we do not cite important papers. We thank many people, in particular M. Azbel, A.T. Bernardes, S. Dasgupta, S. Galam, H.J. Herrmann, N. Jan, H. Milbrodt, T.J.P. Penna, S. Solomon, D.Sornette, Y.C. Zhang and R. Zorzenon dos Santos, for helpful discussion over many years, J.S. de Sa Martins and D. Sornette for a critical reading of the manuscript. Many of our figures were supplied by T.J.P. Penna, A.R. de Lima, J.C. Sartorelli, R.D. Pinto, A.Racco and M. Argollo de Menezes. Most of the book was written while the authors met in Niterói and also at Laboratoire PMMH, Ecole de Physique et Chemie Industrielles, Paris.

Niterói and Köln, December 1998

S.Moss de Oliveira
P.M.C. de Oliveira
D.Stauffer

Instituto de Física, Universidade Federal Fluminense, Av.Litorânea s/n
Boa Viagem, 24210-340 Niterói RJ, Brazil (for SMO and PMCO).

Institute for Theoretical Physics, Cologne University, 50923 Köln, Germany (for DS).

Contents

Part I
Biological Ageing

1 Azbel Phenomenology and Simple Simulations

1.1 Introduction

Even an author who complains about the age of her coauthors is likely to reach their age; and finally we will all die. Death is unavoidable also for animals and plants as well as bacteria (provided we define the splitting of a bacterial cell into two cells as the death of the parent cell). And most olympic records are held by athletes younger than the present authors. Thus physical fitness (as opposed to scientific wisdom) decreases with age during human adulthood, while the probability q to die within the next year increases. For computer simulations, the latter effect, the increase of mortality q with age a, is particularly suitable to define ageing, compared e.g. to studies of beauty.

There is no consensus why we age and die (McCLEARN 1997; JOHNSON 1997; CARLSON and RILEY 1998; MASORO and BERGAMINI 1997; GRIMLEY EVANS et al. 1997; CRISTOFALO and CREPALDI 1998; LIN, SEROUDE and BENZER 1998). Is there a programmed cell death due to telomere loss (HOLBROOK et al 1996; see also OZAWA 1997, WEALE 1997, RUBIN 1997, BODNAR et al. 1998, SEDIVY 1998) which switches our organs off in old age ? Is wear and tear the crucial effect, for insect wings as well as athlete's bones ? Do oxygen radicals and chemical reactions endanger our DNA such that we have accumulated too many errors in our genes when more than a dozen oxygen molecules per body atom (AZBEL 1994; WALLACE 1997; BECKMAN and AMES 1998; PEREZ-CAMPO et al 1998; but see KELLER and GENOUD 1997) were used up in our lifes ? Do we die because at scientific conferences the chairperson does not give us enough time to speak (Azbel, public communication, StatPhys 20, Paris July 1998)? WACHTER and FINCH 1997 collected a dozen review articles on the different aspects of ageing. We emphasize here Monte Carlo simulations for the mutation accumulation hypothesis (ROSE 1991; PARTRIDGE and BARTON 1993, CHARLESWORTH 1994) of population genetics. This theory says that during the evolution of a species many bad hereditary mutations (genetic diseases) happened and were given on to the offspring. Any new mutation may affect any age with equal probability; however, mutations killing us before we get children cannot be given on to later generations, while genetic diseases killing us in old age and unnoticeable before may stay in the population. Thus, in the equilibrium between Darwinian selection of the fittest (which should weed out bad mutations) and new mutations, the pressure of new mutations affects all ages equally, while the selection pressure is much stronger before reproduction sets in. Therefore, in this equilibrium not dissimilar to the one between energy and entropy of statistical physics, the probability of mutations affecting age a to be found in the stationary population is much higher for old than for young age a. We review in the following section the experimental mortality curves $q(a)$ and their analysis by AZBEL (1996, 1997) and some earlier Monte Carlo simulations of models which now seem outdated. Chapter 2 then deals with simulations of the Penna bit-string model (PENNA 1995) and can be regarded as the core of this book.

Finally a historical note: Already four decades ago, the famous physicist Leo Szilard contributed a theory of biological ageing (SZILARD 1959).

1.2 Gompertz-Azbel Phenomenology

Fig.1.1 shows changes in ageing in Germany over more than one century: When adulthood is reached and children's diseases are overcome, the mortality $q(a)$, i.e. the probability to die within the next year, increases exponentially with age a and thus corresponds to a straight line in this semi-logarithmic plot:

$$q(a) \propto \exp(ba) \qquad (1.1)$$

where b is of the order of 0.1/year. This exponential increase was already pointed out in the 19th century by Gompertz for humans (and more recently also for automobiles: page 28 in WACHTER and FINCH 1997). Of course, if q is defined as the probability to die within the next time interval, it cannot become larger than unity and then the Gompertz law (1.1) becomes wrong for old age (older than listed in typical human population tables). A better definition (MUELLER, NUSBAUM and ROSE 1995) lets this time interval go to zero and thus deals with a derivative:

$$q(a) = -d\ln[N(a)]/da \simeq \ln[N(a)/N(a+1)] \quad , \qquad (1.2)$$

where $N(a)$ is the number of individuals with age a in the population. With this latter definition, used already for Fig.1.1, the mortality can go to infinity for $a \to \infty$ as required by the Gompertz law (1.1). (Even better would be to identify $\ln[N(a)/N(a+1)]$ with $q(a+1/2)$ but this trivial shift would not change the curve shape.)

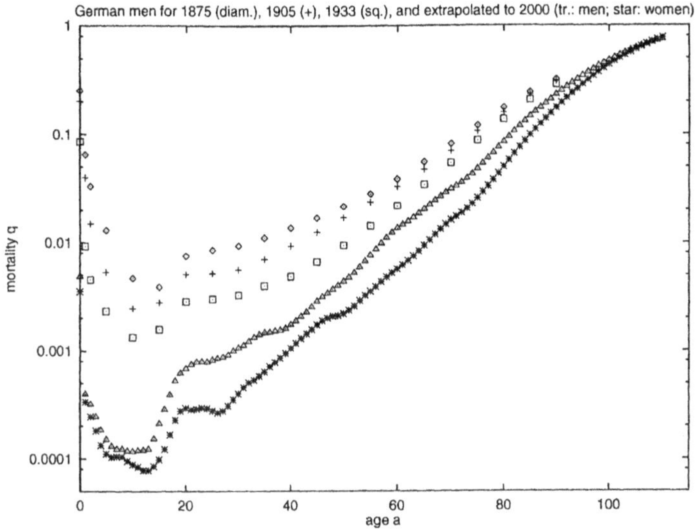

Fig.1.1 - Male mortality in Germany in the past, as well as extrapolated male and female mortalities for the year 2000. The latter data (stars and triangles, revised according to eq.(1.2)) agree well with Japan for 1990. The Gompertz law corresponds to straight lines in this semilogarithmic plot.

AZBEL(1996) (see also STREHLER and MILDVAN 1960) noticed a rather universal law applying to many human populations of Japan, Sweden, Germany during the past 200 years, and also with different parameters to mediterranean fruit flies (the only species for which observations with $> 10^6$ individuals seem published; see AUSTAD 1997 for a comparison of mammals):

$$q(a)/b \propto \exp(b(a - X)) \tag{1.3}$$

with a proportionality factor $A = 11 \pm 2$. Here the characteristic age X is the same for the whole species, i.e. for Sweden as well as for Japan, and for the 18th as well as the 20th century. We call the ratio q/b the dimensionless mortality since the choice of the time unit (years for humans, days for medflies) cancels out. Thus the Azbel law states that this dimensionless mortality q/b reaches a value of the order of ten when humans reach the characteristic age $a = X = 103 \pm 1$ years. Therefore any statistics for $a > X$ is very difficult to get, but X is not a true maximum age. Indeed, humans were reliably reported to live until the age of 122 years. The same law with different X seems to be valid partially for other animals. A more general universality law, which follows from eq.(1.3) where this equation is valid, but also holds for younger age where eq.(1.3) is no longer valid, is obtained when plotting the survival probability against the mean life expectation (Azbel 1999), for humans as well as flies.

[In order to obtain eq.(1.3), Azbel followed the assertion that every death rate has its Gompertz region, where the logarithm of the mortality rate is close to its linear regression

$$\ln q(a) = c^* + ba \quad . \tag{1.4a}$$

Fitting the mortality curves of demographic data Azbel determined the values of c^* and b. He used in fact the dimensionless death rate:

$$\ln[q(a)/b] = c + ba \quad ; \quad c = c^* - \ln b \quad . \tag{1.4b}$$

From each mortality curve he obtained a pair (c, b), and showed that the points determined by c and b follow a straight line, and so can be related by

$$c = \ln A - bX \quad , \tag{1.4c}$$

giving rise to eq.(1.3). It is important to note that the human data used by Azbel refer to different countries and different centuries, giving a whole set of values (c, b). It means that even for a homogeneous stable population, there are variations in the population mortalities, from one year to the other and from one country to the other. However, if all the data refer to one specific species, all pairs (c, b) fall on the same straight line given by eq.(1.4c). Figures 2.4.3 and 2.4.5 later will show simulations analyzed in this way.]

The progress in human living conditions means that the slope b of the Gompertz law increased from one century to the next (if we ignore Hitler's war, not described by the Gompertz law), but that the genetically determined X remains constant over millenia. Thus the mortality for young adults goes down with improved health care, but that at age X remains stable. One may question if these man-made changes of b should be described by genetic population dynamics, but Monte Carlo simulations of RACCO, ARGOLLO DE MENEZES and PENNA (1998) using the Penna model of the next chapter 2 gave agreement with the Azbel universality law (1.3); different civilizations simply corresponded to different random numbers in these simulations.

If Azbel's observation is correct and applies also to the future, then the medical progress would lead to further increases in b but not in X. In this case we would eventually nearly all die at the characteristic age X but be very strong and healthy up to then, quite similar to the catastrophic senescence of salmon discussed in the next section and in the next chapter ("compression of mortality", FRIES 1980). This would correspond to the ideal of ageing therapy to add life to years, not years to life. (Wilmoth in WACHTER and FINCH 1997 summarizes other indications in favor and against such a limit.)

The above law is valid for homogeneous populations having the same Gompertz slope b for all families. Azbel also emphasizes populations with a whole distribution $p(b)$ of Gompertz slopes b, with $\int p(b)db = 1$. For ages a below the characteristic age X, the dimensionless mortality is higher for small b while for $a > X$ it is higher for large b. Thus for very old ages with $b(a - X) \gg 1$ the rare survivors with small b in their genes may have a lower mortality than the many "average" people who died before the characteristic age X. Indeed, the better health of these oldest old has been noticed for humans (PERLS 1995) and was particularly drastic for fruit flies (CAREY et al 1992, CURTSINGER et al 1992), where the mortality decreased with age after some minimum age. However, for humans we are not aware of reliable statistics justifying the assertion on page 14 in WACHTER and FINCH 1997: "Humans who make it to 110 years of age appear to have truly better survival rates than those who make it to 95 or 100", or on page 122: "beyond 85 years, the mortality rate stops increasing exponentially and becomes constant, or actually decreases." The data of Fig.1.1 instead suggest that human mortality of the oldest old increases linearly (PERLS 1995) instead of exponentially with age, in contrast to "the rate of increase slows down" claimed on p.18 of WACHTER and FINCH 1997. Fig.3A of VAUPEL et al (1998) is restricted to women above 80, its change in slope may merely be due to female mortality catching up with male mortality between 80 and 100 years and thus increasing there stronger than the Gompertz extrapolation from younger ages. Fig.4 in THATCHER, KANNISTO and VAUPEL (1998) gives for male mortality a good impression of reality.

Figure 1.2 shows a simple example where the population starts with 10^5 individuals of a "normal" mortality 0.0001 exp($0.1a$) and in addition 10^4 fitter individuals of mortality 0.0001 exp($0.07a$). The numbers of survivors (normalized by 10^5) of the two groups are shown as squares (minority) and diamonds (majority), and the total mortality q as plus signs. Once the majority has died off and the originally smaller group of fitter individuals dominates in the population, the mortality reaches a maximum and then decays, before (not shown) it starts to increase again with the smaller Gompertz slope $b = 0.07$ instead of the earlier $b = 0.1$. Of course, this figure is a mathematical triviality and based on the fact that the sum of exponential functions is no longer another exponential function. Thus with suitable parameters one can get a mortality decreasing with increasing age. The figure here is presented to motivate the more complicated Azbel description explained now.

More generally (AZBEL 1996, 1997; KOLTOVER 1997), the behaviour for

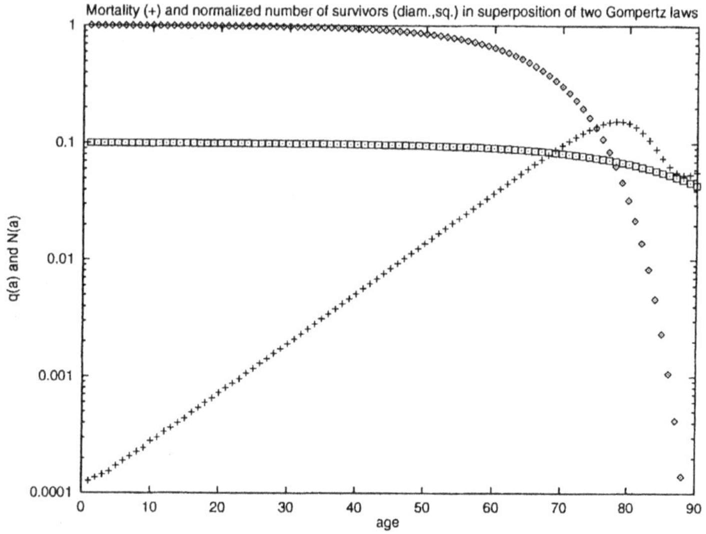

Fig.1.2 - Example of how the superposition of two Gompertz laws of different b (squares and diamonds) can lead to a maximum in the combined mortality q (+). The increase of q beyond the age of 90 is not shown.

these oldest old, $a - X \gg 1/b$, can be described mathematically provided we know whether the distribution $p(b)$ remains positive down to $b \to 0$, or is zero for all b below a positive lower limit b_0. Such a distribution down to $b \to 0$ could have a positive limit $p(0)$ there and jump to zero for negative b. This case is assumed implicitely by Azbel. Or $p(b \to 0)$ could approach zero continuously, or even with a continuous slope, and stay at zero for negative b. Similarly, in the case of a positive minimum value b_0 , the distribution could approach a finite value $p(b_0) > 0$ there and jump to zero for smaller b; or it could be smoother. Fig.1.3 shows schematically these possibilities. We now compute approximately the mortality resulting from these possibilities in a unified calculation, using the Gamma function integral

$$\int_0^\infty x^k e^{-cx} dx = k!\, c^{-k-1} \quad .$$

Let us assume

$$p(b \to b_0) \propto (b - b_0)^k; \quad p(0 < b < b_0) = 0 \tag{1.5}$$

where b_0 is either zero or positive, and k is some positive exponent. A particularly simple case is $k = 0$ where we then have either $p(b = 0) = 0$ (for

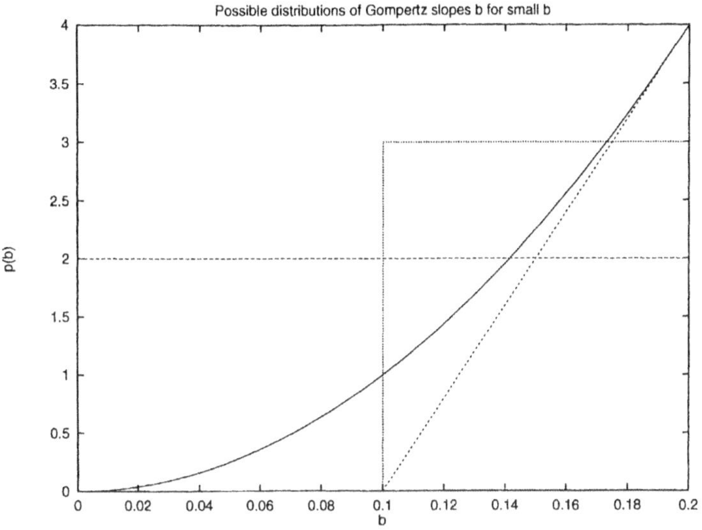

Fig.1.3 - Illustration of eq.(1.5): The distribution of Gompertz slopes b may extend down to $b = 0$ (solid curve and horizontal dashed line) or only down to some positive b_0, here taken as 0.1. In both cases the distribution may approach a positive constant for $b \to b_0^+$ (dotted and horizontal dashed line), or may go to zero continuously with a power law of exponent k (here $k = 1$ for sloped straight line and $k = 2$ for solid curve).

$b_0 > 0$) or $p(b = 0) > 0$ (for $b_0 = 0$). Now we approximate for sufficiently large ages a beyond X the total population $N(a)$, which is a superposition of the populations $n_b(a)$ for each fixed b.

From eqs.(1.2, 1.3) we have $-(1/b) d \ln n_b(a)/da = q(a)/b = Ae^{b(a-X)}$ and thus

$$n_b(a) \propto \exp(-Ae^{b(a-X)}) \quad .$$

Now for large a, using $\exp(e^{\text{const}+\epsilon}) \simeq \exp(e^{\text{const}}) \exp(\epsilon e^{\text{const}})$ for $|\epsilon| \ll 1$ and the abbreviation $B_a = Ae^{b_0(a-X)} \gg 1$:

$$N(a) = \int_0^\infty p(b) n_b(a) db \propto \int_0^\infty p(b) \exp(-Ae^{b(a-X)}) db$$

$$= \int_0^\infty p(b) \exp(-B_a) \exp[-(b - b_0)(a - X) B_a] db$$

$$\propto \exp(-B_a) \int_{b_0}^{\infty} (b - b_0)^k \exp[-(b - b_0)(a - X)B_a]db$$

$$= k! \exp(-B_a)/[B_a(a - X)]^{k+1} \quad .$$

Thus if $b_0 = 0$ we have $N(a) \propto (a - X)^{-1-k}$ while for positive b_0, $N(a)$ decays asymptotically as $\exp(-B_a)$ apart from less important pre-exponential factors. For the mortality $q(a) = -d \ln N(a)/da$ this means

$$q \propto 1/(a - X) \quad \text{for } b_0 = 0 \tag{1.6a}$$

$$q \propto b_0 \exp(b_0(a - X)) \quad \text{for } b_0 > 0 \tag{1.6b}$$

for sufficiently old age $a > X$. AZBEL (1996) fitted for 100 day old medflies the number of very old survivors to a power law decay, eq.(1.6a), with $b_0 = k = 0$. For human beings, no such decrease of the mortality as in fig. 1.2 or eq.(1.6a) was seen for old age in fig.1.1 or analogous Japanese or Swedish data; thus humans may have a much more homogeneous population (all b near a positive b_0) than medflies had ($b_0 = 0$) in the 1992 experiments of Carey et al and Curtsinger et al.

We refer to AZBEL (1997, 1998) for further discussion of the degree of inhomogeneity and its changes with age for human populations. An unsolved question is how to avoid in evolutionary genetics that the fitter minority (smaller b) after many generations replace the less fit majority, just as 10^5 years ago homo sapiens replaced (with exceptions in a nearby capital) Neanderthal man because we (again with exceptions) eat less. Chapter 2 will present the Penna model of 1994 which seems to be the first Monte Carlo model giving the exponential increase of mortality with age (see also MUELLER and ROSE 1996, KOLTOVER 1997 and PLETCHER and CURTSINGER 1998). We will not discuss there other (no-ageing) aspects of population dynamics for which we refer to VANDEWALLE and AUSLOOS (1995).

1.3 Monte Carlo Simulations with Few Ages

The first Monte Carlo simulation of ageing known to us (HOFESTADT 1985) was not adequately published, and thus a continuous flow of Monte Carlo simulations started only following the analytically treated mathematical model in the review article of PARTRIDGE and BARTON (1993). In all these Monte Carlo simulations, one produces random numbers z equally distributed between 0 and 1, from roulette tables or (faster) by computer. These papers were already reviewed by STAUFFER (1994), and their contents or that of this rather condensed section 1.3 is not needed to understand the later simulations based on the multi-age model described in the next chapter 2.

Various papers followed Partridge and Barton and assumed only two ageing intervals: youth and adulthood. The juvenile survival probability from birth (age = 0) to the threshold of adulthood (age = 1) is called J, and that for adult life from age 1 to age 2 is called A. Ageing thus means simply $J > A$. The survivors produce m_1 offspring at age 1, and m_2 offspring at age 2, after which they all die. In each age interval one suffers from M deleterious mutations, where M can be different for youth and adulthood. Too much beer in the youth can be fun but may reduce adult survival; thus Partridge and Barton assumed

$$J + A^x = 1, \quad x = 4$$

to balance the opposing demands of youth and adulthood. A growing population requires $m_1 J + m_2 JA > 1$, which means at least one child per individual.

STAUFFER and JAN (1993) criticized the explanation of PARTRIDGE and BARTON (1993) for the catastrophic senescence of Pacific Salmon , that A jumps to zero discontinuously if there are too many mutations. This fish, long before Lorena Bobbit, warned against the dangers of marriage, since soon after giving birth the parents die. Stauffer and Jan found instead a smooth decay of the survival probabilities proportional to $\exp(-\epsilon M)$ for M somatic mutations per age interval, when each mutation reduces the survival rates by $\exp(-\epsilon)$. (Somatic mutations, in contrast to hereditary ones, are not given on to the offspring .) This result seems trivial if each individual has exactly M mutations, but remains valid if the number of mutations is Poisson distributed with a mean of M. If the number of mutations instead is exponentially distributed with average M, then the reduction factors for the survival probabilities are $1/(1 + \epsilon M)$, again smooth functions without any jumps. These formulas were confirmed by Monte Carlo simulations.

JAN (1994) found that one merely has to put $x = 1$ into the above equation to find the desired catastrophic senescence for salmon. Why this choice of x was made by this fish species is another question; chapter 2 will give a much more intuitive explanation similar to setting $m_2 = 0$ in the above model. For then the condition $m_1 J + m_2 JA > 1$ is fulfilled best by $A = 0$ and $J = 1$, instead of taking both J and A to be smaller than unity.

RAY (1994) checked by analytical theory and by computer simulation that a stationary distribution of young versus old individuals and their survival rates is reached as a function of time t not exponentially in time, as is customary in many models, but with a $1/t$ law as for self-organized critical phenomena .

If instead of somatic mutations one uses hereditary mutations given on to the offspring, then they cannot all be detrimental. DASGUPTA (1994) and

VOLLMAR and DASGUPTA (1994) found stable populations if one has a mixture of positive and negative mutations, both giving about the same absolute change in the survival rates, but with opposite signs. Negative mutations alone lead to mutational meltdown (LYNCH and GABRIEL 1990): The whole population finally dies out even if the birth rate is fixed at an arbitrarily high value. This mutational meltdown is seen particularly easy in a model with only one age interval, where the full program published by BERNARDES and STAUFFER (1995) takes only a few lines: With a detrimental effect ϵ distributed homogeneously within a small interval, even an initially rapidly growing population will finally die out; and this meltdown does not only occur for small populations as asserted by CHARLESWORTH, MORGAN and CHARLESWORTH (1992/93). For further aspects of mutational meltdown we refer to appendix 3 or to the review of BERNARDES (1996a). Mutational meltdown from deleterious hereditary mutations can be avoided, if instead of a homogenous distribution of detrimental effects one has a discrete one, like in a Poisson distribution of mutations where zero effect occurs with a finite probability, or when half of the offspring has zero mutations and half of them two. (If mutational meltdown happens it is eventually unavoidable like natural death; if instead a population dies out because insufficient birth rates etc, it could be avoided by better parameters, like if two male authors shoot each other because of a fight over one woman.)

HEUMANN and HOTZEL (1995) showed that modifications of this model allow for a description of the "dauer" (= permanent) state of certain animals, like hibernation. In this state they consume very little resources. A repetition of bad living conditions like draughts led the species to extinction except if it had the dauer state.

DASGUPTA (1994) also looked at finite-size effects and at a diffusion model for ageing: The individual's J and A values are regarded as $x-$ and $y-$ coordinates in a plane where mutations produce a biased random walk with repelling boundary conditions. There is no need for an a priory asymmetry of young versus old age (introduced by $J + A^4 = 1$ above) for the Dasgupta model to give ageing in the sense of $J > A$. A generalization from two to four age intervals (HEUMANN and HOTZEL 1995) gave a drastic increase of mortality with age, as in reality, but also showed the computational difficulties in this type of approach, which were avoided by the Penna model (PENNA 1995) discussed in the following chapter.

Similarly, PLETCHER and CURTSINGER (1998), continuing MUELLER and ROSE (1996), take the genome as a vector containing ten real numbers $q(a)$ for ages $a = 1$ to 10, each representing the probability to die at that age. A hereditary mutation affects one randomly selected age a by increasing its mortality $q(a)$ by 0.05 (with 99 percent probability), or decreasing it by the same amount (with one percent probability). Together with assumptions that some mutations are good for one age and bad for another age, in special cases instead of the rapid increase of the mortality with age, a plateau was found for $q < 1$. This is perhaps relevant for the unusual behaviour of the oldest old among flies, CAREY et al 1992, CURTSINGER et al 1992.

2 The Bit-String Penna Model

2.1 The Asexual Version

The Bit-String model of life history was introduced by PENNA (1995) and is now by far the most widely used Monte Carlo simulation technique to predict many of the features related to biological ageing ; for recent short reviews see MOSS DE OLIVEIRA (1998) and PENNA et al. (1998). It is based on the mutation accumulation theory already explained in the previous chapter. One well known feature of ageing is that old people can easily describe details of their infancy,but hardly remember what happened 24 hours ago. For two of the authors of this book the main feature of the Penna model is that it is extremely appropriated to numerical simulations, something that they are used to since a long, long time. The reason for this computational advantage is that one individual's hereditary properties are completely described by its genetic code. The genotype of an organism is coded within its cell nuclei in strings called chromosomes . The chromosomes are formed by chemical compounds named genes, which contain all inherited characterics of the organism. The collection of genes present in each of our cells is called genome .

In the asexual version of the Penna Model the genome of each individual is represented by a computer word of 32 bits. It is assumed that each bit corresponds to one "year" in the individual lifetime, and consequently each individual can live at most 32 years. The limit 32 is convenient for computer implementation, but any other value can be used instead. A bit set to one means that the individual will suffer from the effects of a deleterious mutation (disease) in that and all following years. Better explained, we start the simulation with $N(0)$ individuals or bit-strings , that can be either randomly chosen or free of mutations. Free of mutations means that all strings contain only zero bits, and randomly chosen means that each string contains zeroes and set bits randomly distributed. The only difference between these two ways of starting the process is the amount of time required to reach a stable population and the behaviour of $N(t)$ during equilibration. One step of the simulation corresponds to reading one bit of all genomes. Whenever a new bit of a given genome is read, we consider that the individual is one "year" older.

The rules for the individual to stay alive are: 1) The number of inherited diseases (set bits) already taken into account until its current age must be lower than a threshold T, the same for the whole population. As an example, an individual with a genome represented by the string 100101.... would die at age 4 if $T = 2$, and at age 6 if $T = 3$. 2) The limits of space and food are taken into account through the Verhulst factor $V = 1 - N(t)/N_{max}$, where

N_{max} is the maximum population size the environment can support (carrying capacity) and $N(t)$ is the current population size. The carrying capacity is usually considered ten times larger than the initial population $N(0)$. At each time step of the simulation a random number between 0 and 1 is generated for each individual and compared with V. If it is greater than V, the individual dies independently of its age or genome. The smaller the population size is, the greater is the probability of any individual to stay alive. The Verhulst factor is of great importance in computer simulations in order to avoid the population to explode, and can be seen as a kind of mean field interaction that couples individuals of different ages. (See RAAB 1998 for criticism of this application of Verhulst factors.) Now, reproduction and mutations: when the individual succeeds in staying alive until the minimum reproduction age R, it generates b offspring at age R and all following years. The offspring genome is a copy of the parent one, except for M randomly chosen mutations introduced at birth. These mutations can be either good or deleterious. A good mutation is obtained when the randomly chosen bit of the parent genome is equal to one, and so is set to zero in the baby genome. The opposite case corresponds to a bad mutation. Summarizing, the Penna Model works with the following parameters:

a) $N(0)$ - initial population;
b) N_{max} - maximum population size;
c) T - limit number of inherited diseases ;
d) R - minimum reproduction age;
e) b - birth rate;
f) M - mutation rate in offspring genome.

While the whole genome is given to the offspring at birth and thus contains the date of the genetic death of that individual, only that part of the genome is visible and active which corresponds to the present age of the individual: Its genetic future is already stored but not yet known. MULLER-HILL (1993) and the 1997 movie Gattaca warned already of the consequences for human society if the whole genetic future becomes known for us.

2.2 Important Remarks

• In living beings the genome size can vary from 10^7 up to 10^9 units, depending on the complexity of the organism. The extremely short size of the 32 bit-string of the Penna model would be totally unrealistic if all our genes were related to life threatening diseases. However, only around 10^4 to 10^5 of the units play a functional role, and only a subgroup of these will give rise to a serious disease at some moment of the individual lifetime. Besides, qualitatively there was not much difference when 32, 64 and 128 bits were taken into account, as obtained by PENNA and STAUFFER (1996). Another point is that although we call genome the bit-string of the model, it is in fact representing the consequences of the errors (diseases) contained in the real genome. The biological motivation is that the Alzheimer disease, for instance, only appears at old ages, although it is present in the genetic code since birth. BERNARDES and STAUFFER (1995) studied an integer version of the Penna model, in which the single bit equal to 0 or 1 is replaced by an integer varying between 0 and 9, and representing several life-threatening inherited diseases . The individual dies if the sum of all integers up to its current age exceeds some threshold. They concluded that this approach consumes more computational time and memory, but does not change significantly any of the results obtained with the original model.

• Although the model allows good and bad mutations, generally only the bad ones are taken into account. Harmful mutations apparently arise at a rather high rate in most organisms, and the frequency of backward mutations (reverse mutations deleting harmful ones), as pointed out by PAMILO, NEI and LI (1987), is 1/100th the frequency of forward mutations. When only bad mutations are considered, if an already set bit is randomly chosen in the parent genome, it remains equal to one in the offspring genome. Although in this case the offspring is always equal or worse fitted than the parent, a stable population is obtained, provided the birth rate is greater than a minimum value. This minimum value will be analytically obtained in section 2.3, for the specific case of Pacific Salmon (and also bamboo) where catastrophic senescence occurs. However, it is important to note that the population is sustained by those cases where no mutation occurs, when an already set bit in the parent genome is chosen. If one decides to adopt a model in which only zero bits are randomly chosen to be mutated, then it will always end up with a null population (see appendix 3). Extinction due to accumulation of deleterious mutations is called mutational meltdown and was first considered by LYNCH and GABRIEL (1990). Details can be found in BERNARDES (1996a) excellent review on Monte Carlo simulations of biological ageing; for more informations about mutational meltdown in two-age models, see BERNARDES and STAUFFER (1995). The unavoidable mutation meltdown, when each generation adds more deleterious mutation to the genome, may not depend on the population size, as obtained by BERNARDES (1995), and has to be distinguished from avoidable extinction due to insufficient birth rates or due to random fluctuations in small systems (PAL 1996).

• Although we have been dealing only with hereditary mutations , *somatic* ones can also be considered. These are mutations that arise during the lifetime of an individual, but are not transmited to the offspring. Skin cancer is a well known example. They have been introduced in the *sexual* version of the Penna Model in order to obtain a greater survival probability for women than for men (section 2.5.3). The importance of somatic mutations for death was already emphasized by CLARK (1996).

• We have pointed out in the beginning of this chapter that the Penna Model is particularly suitable for computer simulations. The main reason is that individuals are represented by computer words that can be manipulated bit by bit through the logical operations **AND**, **OR**, **XOR** (exclusive OR) and **NOT**. These operations work in the following way:

0 **AND** 0 = 0	0 **OR** 0 = 0	0 **XOR** 0 = 0	**NOT** 0 = 1
1 **AND** 1 = 1	1 **OR** 1 = 1	1 **XOR** 1 = 0	**NOT** 1 = 0
1 **AND** 0 = 0	1 **OR** 0 = 1	1 **XOR** 0 = 1	
0 **AND** 1 = 0	0 **OR** 1 = 1	0 **XOR** 1 = 1	

In order to obtain a bad mutation, one needs to do an OR operation between the parent genome and an auxiliary computer word that contains a single bit set in the position where the mutation will occur. Let's supppose, for example, that a parent genome has four bits and you're going to introduce one randomly chosen mutation on the second bit (from left to right in written representation). The auxiliary word in this case is **0100**. Considering, for example, that **0010** is the parent genome, the offspring one will be **0100** OR **0010** = **0110**, now with a new bad mutation at position two if compared with the parent genome. It is easy to see that if the randomly chosen bit of the parent genome is already set, no mutation occurs. Alternatively, the XOR operation holds for the case of good and bad mutations. In order to count the number of diseases accumulated in the genome the AND operation is used with properly chosen auxiliary words.

The FORTRAN program of the asexual bit-string model is already published by PENNA and STAUFFER (1995), and the sexual version can be found in appendix 2, together with an improved asexual one. In this improved version there is an extra word for each individual i, $DATA(i)$, that stores all informations about it: *age at bits 0 to 5* and *counted mutations at bits 6 to 11*. In order to check the individual age the instruction [AGE = 63 AND DATA(i)] is used (63 is a bit-string with the 6 first bits equal to one and the others equal to zero). To count the number of mutations the following instructions are performed:

NMUT = 63 AND [SHIFT (DATA(i),−6)] (reads the amount of mutations already stored in the word DATA, after a shift of six bits),

N = 1 AND [SHIFT (GEN(i), −AGE] (checks if the bit of the genome corresponding to the current age is set or not),

IF N = 1, NMUT = NMUT + 1 (increases the stored number of mutations).

There has been a proposal by one of the authors to adopt the C language in the appendix, but it was refused by another one, who is also known as Tyrannosaurus Imperator. However, the C version can be sent, if requested, by e-mail. Using this bit-string technique it has been possible to simulate population sizes comparable to those found in Nature, as will be shown in the next section.

2.3 Applications of the Asexual Penna Model

We are now going to present many results obtained through this model in order to show that they are compatible with biological facts. Different aspects of the dynamical process will appear, as well as its capacity to reproduce many of the known features presented in real populations. Even chaotic behaviour like in the logistic map, to be discussed in chapter 3 for leaky faucets, heartbeats and bird songs, was found in the Penna model when the birth rate was high and the minimum age of reproduction was lower than the threshold T (BERNARDES, MOREIRA and CASTRO-E-SILVA, 1998). However, its most basic and important aspect is the appearence of *ageing*, defined as *a decrease in the survival probability with age*. In all applications that follow *only bad mutations are considered*, except for the cases where good ones are explicitly mentioned. The main results of the model as the survival rates, for example, are the same, if good mutations are included. The total population, however, is greater in this case.

A smart reader may ask if presenting the main features of the Penna model in sub-sections, like we are going to do, is not a trick to force her/him to read the subsequent pages of the book; the answer is yes, it is! However, a summary of these main features is presented in the last section of this chapter (of course readers skipping the next pages will also age faster).

2.3.1 Fishing

Around 1994 a great deal of information concerning the decrease of world fish stocks and the consequences of abusive fishing was published. The sudden disappearence of Northern Cod (ROSE 1993) and other fish along the Atlantic coast of North America, in particular of Newfoundland, is a good example. It was also pointed out that the disappearence of some kinds of fish that usually eat algae is also responsible for the partial extinction of caribbean coral reefs, that are being covered by these algae and consequently dying. However, can fishing alone be responsible for such a catastrophe? Is it possible to destroy a stable population simply through fishing? Motivated by these questions MOSS DE OLIVEIRA, PENNA and STAUFFER (1995) included fishing in the Penna model. After 1000 steps, when a stable population was already obtained, a rate of fishing was introduced and kept fixed until step 1200, when a small extra fishing started. The result can be seen in Fig.2.3.1.

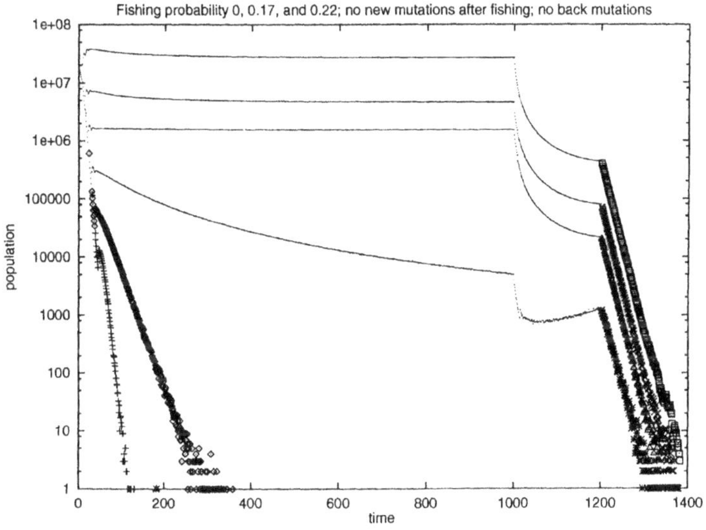

Fig.2.3.1 - Development of population in time. Starting with $N(0) = 20$ million babies, increasing to 38 million, stabilizing without fishing near 27 million. Thereafter fishing first reduces, then kills the whole population. Parameters: $N_{max} = 10N(0)$, $T = 3$, $R = 7$, $b = 1$ and $M = 1$. The top curve shows the total population, then (from top to bottom) those of ages 1, 8, 16, 24 and 32.

In fact, there are many interesting features in this figure to be analysed. The first one to be noticed is the ageing effect, that can be seen from the fact that after 400 steps there are no fish older than 24 years alive (we will maintain the notation "year" as the difference in time between two consecutive steps of the simulation). As explained before, there are two ways of starting the simulation: with all the genomes free of mutations or with a population where each genome contains half of the bits set in random positions. In both cases the population decays in the very beginning due to the Verhulst factor, and for the random case it decays even more because of the harmful mutations in the genomes. As soon as the first reproduction occurs, the population increases again and as a result, one needs to wait for a transient period to obtain a constant population with a death rate compensated by an equivalent birth rate. This transient period generally corresponds to 100 steps, and can be noticed at the most left part of the curves in Fig.2.3.1. However, a constant population does not mean a *stable* population , which happens when *the number of individuals at any given age is constant in time.* For this reason, older animals disappear only after several hundred steps of the simulations. The time required to obtain a stable population must be measured in generations. The same is true for all the genetic features we are going to see, that are only fixed in the population after many generations. The simple fact that older individuals disappear but not the younger is a signature of ageing, and a consequence of the dynamics that kills individuals after T mutations. It is easy to understand what happens: those that have a great number of bits set in the initial part of the genome are not able to reach the minimum reproduction age R and disappear from the population. The only way to survive until reproduction and so to perpetuate the species is to have a genome as clean as possible until age R. However, mutations are there! Where? Accumulated at the end part of the genomes , decreasing the probability of survival with age.

Returning to fishing, it can be noticed that after the introduction of a 17% rate of fishing at step 1000, the total population decays but would have stabilized again if an extra fishing of 5% were not introduced at step 1200, leading to extinction. So it is possible to conclude that fishing works like a first order phase transition: if a threshold is slightly surpassed, the whole population disappears. In this same paper (MOSS DE OLIVEIRA, PENNA and STAUFFER 1995) the reader can observe what happens if only fish older than reproduction age are caught and also how the sustainable fishing rate varies with the birth rate. The failure of a mean field equation to describe the process can also be found there.

2.3.2 Catastrophic Senescence

Catastrophic Senescence means death soon after reproduction, but not immediately after as a consequence of some complication in the very moment of reproduction, like sexual cannibalism by some female (VANDEWALLE 1997). The best known examples of catastrophic senescence are the salmon and the mayfly. In the salmon case, one could ask if it is not a consequence of the hard travel this species is forced to do in order to reproduce in sweet water instead of in the ocean, its natural habitat. When it enters fresh water it stops eating, and some kinds such as the King Salmon travel 3200 km up river to reach their spawning place. So the question that arises is equivalent to that of "which comes first, the chicken or the egg?", that in this case can be translated as "does the salmon reproduce only once because it dies or does it die because it reproduces only once?". The answer was given by PENNA, MOSS DE OLIVEIRA and STAUFFER (1995) and repeated qualitatively by S.Tuljapurkar on page 70 of WACHTER and FINCH (1997).

Before presenting the key to the salmon effect, and also to keep the reader in suspense, let us introduce the survival probability or survival rate . For an already stable population it is given by the ratio between the number of individuals with age $a+1$ and the number of individuals with age a:

$$S(a) = \frac{N(a+1)}{N(a)} \quad . \tag{2.3.1}$$

The normalized survival rate is obtained by dividing all the survival rates by the first one, that is, by the survival rate from age 0 to age 1. The advantage of this normalization is to eliminate the Verhulst factor, and to obtain, for an already stable population, results that do not depend on the size of the population but only on its genetic charge. Fig.2.3.2a shows the normalized survival rates as a function of age, where a maximum reproduction age R_{max} was also introduced. The curves were obtained for the following common parameters: $N_{max}/N(0) = 10$, $T = 1$, $b = 1$ and $M = 1$. x correspond to reproduction from age 6 to 12; squares from 10 to 12; diamonds from 6 to 32 and crosses to reproduction from 10 to 32. These curves were obtained in a few hours on a Pentium, for several 10^5 individuals.

The first important effect to be noticed is that *ageing starts with reproduction*: the survival rate decays as soon as reproduction age is reached (x and diamonds start to decay before crosses and squares). Another important feature is that (after many generations, when a stable population has already been achieved) *there are no more individuals alive older than the maximum reproduction age* (x and squares). The catastrophic senescence corresponds to the case in which only one reproduction age exists (Fig.2.3.2b) and is the same for all individuals. The only explanation we have for these effects is the Darwinian one: individuals must stay alive in order to reproduce and perpetuate the species. If they can no longer generate offspring but remain in the population, they will compete with the younger ones for food and space. So this is the key we mentioned before for the salmon problem: it dies because it reproduces only once. It is important to note that it naturally comes out through the Penna model dynamics: above the maximum reproduction age,

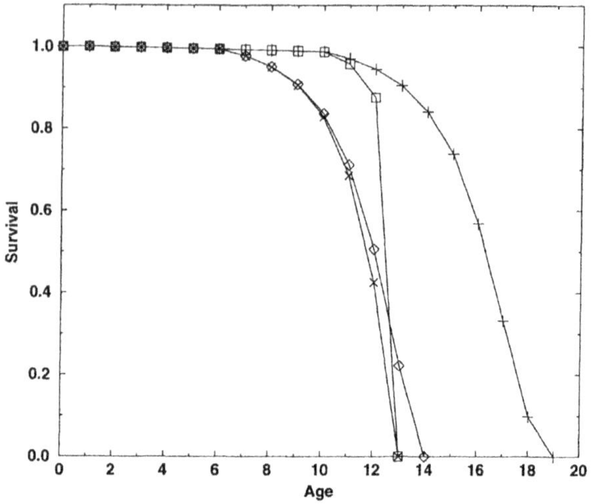

Fig.2.3.2a - Normalized survival rates as a function of age. Each curve corresponds to a given period of reproduction: x from 6 to 12; squares from 10 to 12; diamonds from 6 to 32 and crosses from 10 to 32.

there is no selection pressure balancing the mutation pressure, and thus all bits get set. More recently, FEHSENFELD et al (1998) introduced a strong form of parental care (see also section 2.3.4) into these semelparous populations (that reproduce only once in life at age R). They assumed that only babies with living mothers may survive, until they reach a given age A_{SPC}. They found, for both sexual and asexual reproductions, that the final survival age of the whole population is pushed from R to $R + A_{SPC}$ when this child protection is considered. As a final remark, if the other parameters of the model are moderately changed (T, M and b) the normalized survival rate is not much affected (although the total number of individuals is sensible to these variations).

Since we have now answered the question posed before, let us leave the reader with another one: if death occurs when reproduction stops, why do women live even longer than men, despite the menopause they are subjected to? Continue to read the book and you will find the solution when sex is finally included in the model.

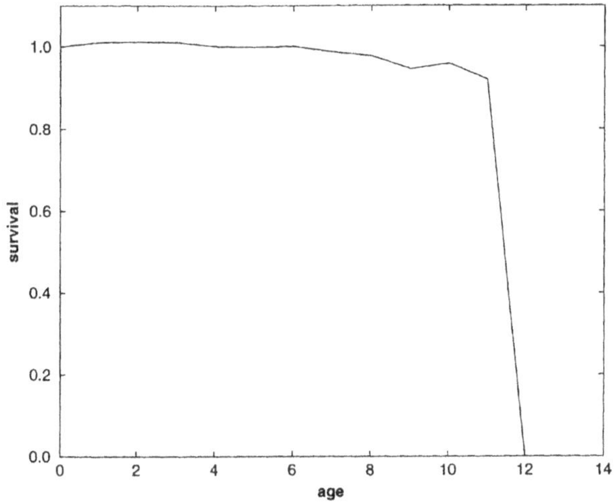

Fig.2.3.2b - Normalized survival rate for only one reproduction age $R = 11$; $T = 3$, $b = 2$ and $M = 1$.

2.3.3 Exact Calculations for Catastrophic Senescence

Analytical calculations for the complete model are possible, although not very simple, and will be mentioned later. In this section we present exact solutions for the case where there is only one reproduction age R, and $M = T = 1$, that is, only one mutation per offspring genome and just one deleterious mutation is enough to kill the individual. Let us assume that the system has already reached an equilibrium state. Since the age distribution and the population are stationary, we can consider the Verhulst factor V as a constant. Let us consider there are N individuals with zero age, and that B is the genotype length, in bits. The probability of a mutation to appear at a given age is $1/B$, for ages below R, since in this simple case ($T = 1$) we can be sure that the parent has no mutation up to age R (or it would have died before giving birth). For ages beyond R the whole parental history must be taken into account, and this will be done later. First ignoring births, we can expect that the number of individuals with age 1 will be:

$$N_1 = NV\left(1 - \frac{1}{B}\right) = NV\left(\frac{B-1}{B}\right) \quad . \tag{2.3.2}$$

Analogously, the number of individuals with age 2 is given by:

$$N_2 = N_1 V\left(1 - \frac{1}{B-1}\right) = NV^2\frac{(B-1)(B-2)}{B(B-1)} \quad . \tag{2.3.3}$$

For general age $a \leq R$, one has:

$$N(a) = NV^a\left(\prod_{i=1}^{a} \frac{B-i}{B-i+1}\right) = NV^a\frac{(B-a)}{B} \quad . \tag{2.3.4}$$

Now, we take into account the reproduction at age R. Since each individual produces b offspring, the number of individuals with age zero will be b times the number corresponding to age R in the previous step:

$$N_0 = bNV^R\frac{(B-R)}{B} \quad . \tag{2.3.5}$$

Using the above equation, with $N = N_0$, it is possible to obtain the stationary state condition:

$$bV^R(B-R)/B = 1 \quad . \tag{2.3.6}$$

As we should expect, this condition does not depend on a but only on the global parameters R, B and b.

For ages $a > R$ we must guarantee that no mutation ocurred also between ages R and a. The probability that no mutation has ocurred until age a, in the $t^{th} = 1, 2,$....ancestor (parent, grand parent, etc.) is, following the same reasoning leading to eq.(2.3.4):

$$\left(\frac{B-a}{B}\right)^t \quad , \tag{2.3.7}$$

and this is a necessary condition for the individual to be still alive, to be fulfilled for all ancestors. However, this term goes to zero as t increases. Thus, in order to have the stationarity condition satisfied, we must have:

$$N(a > R) = 0 \quad , \tag{2.3.8}$$

corresponding to the catastrophic senescence effect. Now it is clear that the only important ingredient for these phenomena is to reproduce only once, all individuals at the same age.

Again using the stationarity state condition (2.3.6) it is easy to obtain V:

$$V = \left(\frac{B}{b(B-R)}\right)^{1/R} \quad . \tag{2.3.9}$$

Now introducing this last result into eq.(2.3.4) we get an expression for the stationary age distribution:

$$N(a) = N\left(\frac{B}{b(B-R)}\right)^{a/R} \frac{B-a}{B} \quad . \tag{2.3.10}$$

Using the definition $V = (1 - N(t)/N_{max})$ we find from eq.(2.3.9) that:

$$N(t) = \sum_{a=1}^{R} N(a) = N_{max}\left[1 - \left(\frac{B}{b(B-R)}\right)^{1/R}\right] \quad . \tag{2.3.11}$$

From this expression it is possible to notice that *the total population is a decreasing function of the reproduction age.* These analytical calculations were performed by PENNA and MOSS DE OLIVEIRA (1995), for a general value of M, presenting an excellent agreement with computer simulation results.

From eq.(2.3.11) it is possible to obtain the minimum birth rate, b_{min}, necessary to avoid population extinction:

$$b_{min} = B/(B-R) \quad . \tag{2.3.12}$$

The dependence of b_{min} with M is also shown in the original paper from Penna and Moss de Oliveira.

As mentioned before, the relevant quantities in ageing studies are the survival rates , which can easily be obtained from eq.(2.3.10):

$$S(a) = \frac{N(a+1)}{N(a)} = \left(\frac{B}{b(B-R)}\right)^{1/R} \frac{B-a-1}{B-a} \quad . \tag{2.3.13}$$

The remarkable agreement between computer experiments and eq.(2.3.13) can be seen in Fig. 2.3.3.

2.3.4 Privileged Conditions

Many conditions can be introduced in the Penna Model in order to privilege groups of individuals and check the evolution of these groups in comparison to the non privileged ones. MOSS DE OLIVEIRA, DE OLIVEIRA and STAUFFER (1995) introduced this kind of approach to obtain the Oldest Old effect, already mentioned in the first chapter. Their purpose was to simulate the fact

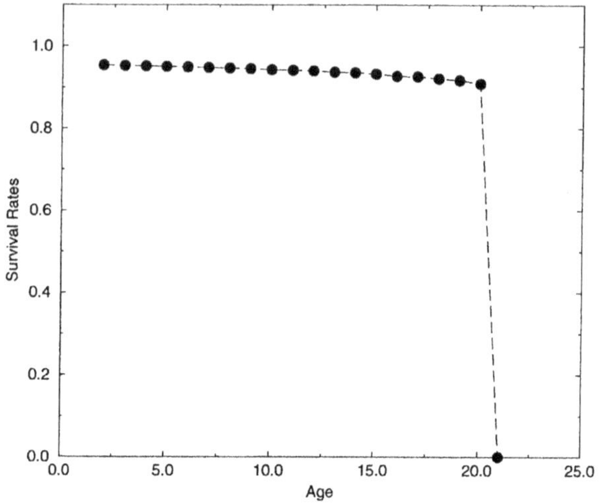

Fig.2.3.3 - Survival rates from analytical calculations (line) and from computer simulations (symbols): $M = 1$, $b = 5$ and $R = 20$.

that these oldest old individuals, generally older than ninety years in case of human beings (PERLS 1995), may present a higher survival rate in this advanced age than that of earlier ages. The authors would have enjoyed a lot if such a phenomenon could have been observed in younger individuals. However, Nature seems to be not so kind, and there is no way to avoid a beer belly, to postpone those terrible wrinkles that allow plastic surgeons to earn lots of money, or to preserve the same creativity in theoretical physics.

The oldest old effect is related to the fact that the average survival probability from ninety to ninety five years may be greater than that from eighty to eighty five, or at least does not decay as fast as for earlier ages. Three different strategies were simulated, but only in two of them a genetic criterion is used to select the privileged individuals. The adopted strategies are:

i) Instead of counting all 1-bits in the genome up to the current age, in order to compare the result with the survival limit T, only the 1-bits corresponding to the last K years of the individual life are taken into account. This privilege holds for all, and means that the body, after K years, has adjusted to an old

sickness and no longer is threatened by it (immunity).
ii) Individuals having a 0-bit at *some particular positions of the genome*, defined a priori, generate offspring that can have, with probability 50%, good and bad mutations. For all other individuals only bad mutations are allowed. The simulation is performed using a fixed word W of 32 bits, defined at the begining of the program, that contains 1 bits at the chosen privileged positions. Only if the parent genome (GEN) presents 0-bits in all those positions, the offspring can have also good mutations (meaning the reversal of bad ones). The test is made through logical operations, comparing [W AND NOT(GEN)] with W. Note that the privilege is now transmitted to the offspring of the particular set of individuals that fit the word W.
iii) The third strategy is equivalent to the previous one, with the advantage that the word W is generated by the dynamics itself, i.e., W is dynamically self organized by considering the overlap between genomes of the oldest individuals in the current population. Because this strategy is more complicated than the second one, but leads to equivalent results (the privilege is the same), we request the reader who is interested in it, to check the original paper for details.

In Fig.2.3.4 we show the results obtained for the second and third strategy, comparing the survival rates for the privileged cases and the non-privileged one. The result obtained with the first strategy is equivalent, although much more pronounced, since all individuals can potentially reach the privilege at old ages.

It is interesting to note that with the two last strategies it was possible to keep two different populations living together, and so to avoid the effect mentioned in the first chapter (Azbel theory) that the most fit group tends to dominate the whole population, leading to the disappearence of weaker ones. More recently DE OLIVEIRA et al (1999) have shown that in fact it is not necessary to include special conditions in the Penna model in order to obtain the oldest old effect. One needs only to adopt suitable parameters, as $T = R = 3$, $M = 4$ and $b = 1$ and large populations. They have also found - DE OLIVEIRA et al (1998), studying the correlation between the genetic death ages of parents and children, that longevity is heritable (for both sexual and asexual populations). Also MUELLER and ROSE (1996) presented Monte Carlo simulations which gave for young and intermediate ages a rapid increase of the mortality, followed by a plateau for old ages; see PLETCHER and CURTSINGER (1998) for some criticism and generalization of this work. In our case of privileged conditions, the non-privileged subpopulation is continuously fed by the privileged individuals (a privileged parent can always give birth to a non-privileged offspring). This kind of strategy also leads to a higher genetic diversity of the population, measured through the

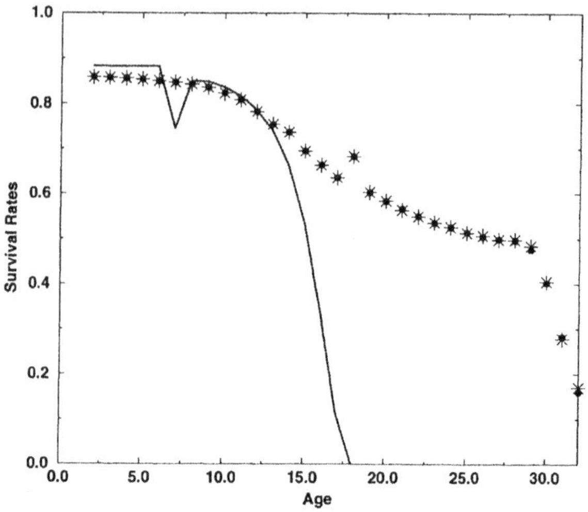

Fig.2.3.4 - Survival rates versus age for the cases where: (a) There are no privileged individuals (dotted line); (b) The privileged set is defined a priori (circles), W containing five 1 bits; (c) The privileged set is dynamically generated by the system itself (stars). Sets b and c are nearly indistinguishable. For all cases $N(0) = 500,000$, $T = 3$, $R = 8$ and $M = 1$.

Hamming distance (number of different bits between the genomes for all pairs of individuals. The result is shown in Fig.2.3.5.

A specific genome has also been used by THOMS, DONAHUE and JAN (1995) in the Penna model with good and bad mutations. Now a pre-defined word W is considered the best genome to survive in a given environment. They start the simulations with half of the individuals having the optimal sequence. One, two or zero mutations are introduced at birth, with equal probability. However, a mutation is considered deleterious only if its corresponding bit is different from that of the optimal sequence. In this way, the randomly selected bit of the parent is always switched to the opposite state (except for the cases where the probability of mutation is equal to zero). The comparison is done just up to the current individual age. This procedure is on average equivalent to the standard Penna model where W = 0. In their simulations Thoms, Donahue

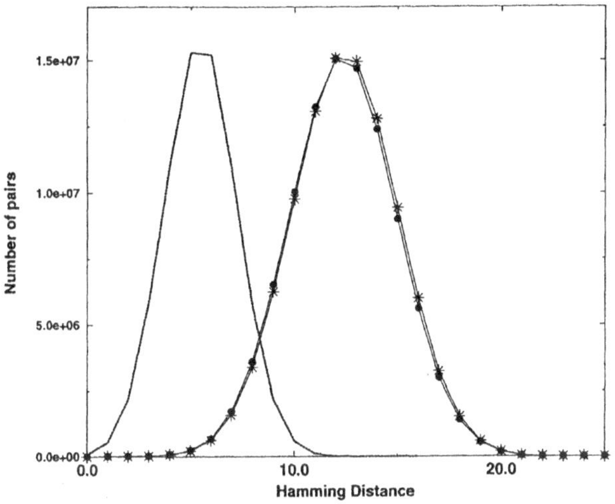

Fig.2.3.5 - Histogram of the Hamming distance for all pairs of individuals; the notation and parameters are the same of Fig.2.3.4. for a smaller initial population.

and Jan admit also somatic mutations , that may appear with probability 0.01 every year. The average number of deleterious mutations per individual is measured, at each time step, and a Fermi type (genetic) death probability p_d is adopted:

$$p_d = \frac{1}{\exp(\tau(\overline{n} - n)) + 1} \quad , \tag{2.3.14}$$

where $\overline{n}$ is the average number of mutations per individual and n is the number of mutations of the individual concerned; τ is set equal to 10, and gives a low survival probability to those individuals with a high number of mutations and a high survival probability to those whose number of mutations is smaller than $\overline{n}$. In this way the lower τ is, the more flexible is the environment. This death probability is then compared with a random number between zero and one. If the random number does not exceed p_d, the individual dies. This procedure replaces the former test of threatening diseases in the original Penna model. A systematic investigation of the differences caused by this smooth Fermi function instead of a sharp cutoff is still missing.

Assuming a specific period of reproduction from age 11 to 25, they observed that the *accumulation of deleterious mutation is about 10 times higher at older ages than that before the reproduction period*; a similar result was also found by STAUFFER (1996, fig.2), using the standard Penna model.

Another interesting situation of privileged conditions in the Penna Model was studied by THOMS et al (1995), by the introduction of parental care. They used the same version of the model explained above, but allowing those individuals whose parent is still alive to escape from the Verhulst factor. They obtained a noticeable increase in life expectancy and in the overall fitness of the population. For a reproduction range from 10 to 25 and a maximum lifespan of 64, they got an average number of mutations for the group of 13-25 without parental care aproximately 70% higher than that with parental care.

The Verhulst factor can also be used to couple different kinds of species, as will be shown below.

2.3.5 Changes in the Verhulst Factor

PUHL, STAUFFER and ROUX (1995) studied the case where both prey and predators are simulated by the Penna Model, assuming two different Verhulst factor, one for prey and another for predators. The survival probability of the predators is assumed to be proportional to the number N_1 of prey, and the death probability of the prey is proportional to the number N_2 of predators. The Verhulst survival factors are given by:

$$V_{prey} = 1 - N_2(t)/N_{max2} \quad ; \quad V_{pred} = N_1(t)/N_{max1} \quad . \qquad (2.3.15)$$

As a result they got a population that strongly oscillates at the beginning, but these oscillations are rapidly damped, although they never disappear. In order to check if the oscillatory behaviour was obtained due to the strong coupling between the two populations, BERNARDES (1996a) divided the survival term in two terms: One due to space restrictions depending on a general parameter for each population, and the other depending on the ratio between the two populations. These terms are:

$$V_{prey} = 1 - (N_1/N_{max1}) - (N_2/N_1) \quad ; \quad V_{pred} = 1 - (N_2/N_{max2}) - (N_2/N_1) \quad . \qquad (2.3.16)$$

With these terms, the oscillations disappeared.

FEINGOLD (1996) applied the Penna Model to the study of Alaska wolves. These animals are currently hunted intensively, and it is estimated that 30% or more of the whole population is killed every year. Wolves are predators which live in packs, and their social structure is extremely important for survival.

However, intense hunting may disrupt this social behaviour, increasing even more their death rate. In order to simulate this effect, Feingold included in the Verhulst factor a social disruption term, together with a hunting rate:

$$V = 1 - N(t)/N_{max} - \text{hunting} - s_d \quad ; \quad s_d = sN(t)/N_{max0} \quad , \tag{2.3.17}$$

where s is an arbitrary constant and N_{max0} is the size of population at equilibrium. This expression describes a social disruption that affects the whole population equally. The simulations were performed with strings of lenght 16. As a result it was obtained that at high hunting rates and high values of the social disruption term, the population does not reach an equilibrium size and asymptotically goes to zero.

CEBRAT and KAKOL (1997) also studied this same problem of social behaviour, and as far as we know their paper is the first one published on the Penna model by non physicists (they belong to the Department of Genetics, Wroclaw University, Poland). Besides hunting, they introduced a social alliance factor - s - instead of the social disruption factor s_d of Feingold into the Verhulst factor: $V\alpha(1 - N/sN_{max})$. The social alliance factor is a constant showing how much the maximal capacity of environment grows with the organization of the population. Many comparisons are made between populations with and without social behaviour, as well as studies about the population evolution depending on when each term of the Verhulst factor is switched on.

More recently, MAKOWIEC (1997) made a study on the effects of the various parameters of the model in the whole population, including the Verhulst factor influence. The special role of social alliances was interpreted as a way to protect young individuals, and an age-dependent Verhulst factor was considered:

$$V_f(t,a) = \left(1 - \frac{N(t)}{N_{max}}\right)^{a/f} \quad ,$$

where f denotes the last protected-favoured age. Results for different values of f are presented as well as comparisons with the traditional Verhulst factor. The effects of a periodically changing Verhulst factor was also studied:

$$N_{max}(t) = N^o_{max}(A + \cos(Bt))/2 \quad ,$$

giving that if $A > 1$ then a stationary population develops depending critically on the value of B.

The Verhulst factor may have an important role in order to simulate the bit-string model on a lattice and study the spatial distribution of the population. In this case, it is necessary to attribute a Verhulst factor (carrying capacity) per site (or per region). Presently, A. O. DE SOUSA and MOSS DE OLIVEIRA

(1998) are investigating the final spatial configuration of a population when two different strategies of child-care are considered. The simulation starts with a random diluted lattice, with one individual at each occupied site. Every individual older than age C has a probability to move to a neighbouring site that is less or equally occupied than its current one. Every site has the same carrying capacity. The child-care strategies are: a) the mother cannot move on the lattice if she has any child younger than C; b) she can move, but keeping children younger than age C with her. Sousa and Moss de Oliveira have obtained rather different final configurations, those corresponding to strategy a) with more empty spaces than in case b).

2.3.6 Strategies for Reproduction

For the readers that have been waiting for sex to appear, we ask for a little bit more of patience, because this section still refers to asexual reproduction . Sex is always a complex subject, and for one of the authors computer simulations are the only available way to understand it. In order to give him some more time, appropriate to his advanced age, let us now better investigate how the range of reproduction and the birth rate affects the population in the Penna model.

It has already been shown (section 2.3.2) through the normalized survival rates that ageing starts with reproduction, that the earlier the minimum reproduction age is, the smaller is the population lifetime, and that soon after reproduction stops the individuals die. These results were confirmed by THOMS et al (1995) using bit-strings of 64 bits. They simulated two different populations, one with a reproduction range from 10 to 25, and the other from 20 to 35. Measuring the number of individuals as a function of age, they observed that the curve representing the older reproductive period has an extended range when compared to the other one. However, *the size of the steady state population decreases as the reproductive range is shifted upward*, while the fitness of the population increases.

ARGOLLO DE MENEZES, RACCO and PENNA (1996) reproduced the observed feature of trees and some species of fish whose fertility increases with age a, resulting in a high maximum lifespan. They adopted three different forms of the birth rate b: $b = b_0$ (constant as in the original model), $b = b_0 * a/32$ (linear increase with age) and $b = b_0 * (a/32)^2$ (parabolic increase). The last two cases give a maximum lifespan around 33% greater than the first one, as can be observed in Fig.2.3.6.

Another interesting strategy of reproduction was studied by ITO (1996a). He assumed that the minimum reproduction age *minage* is variable from animal

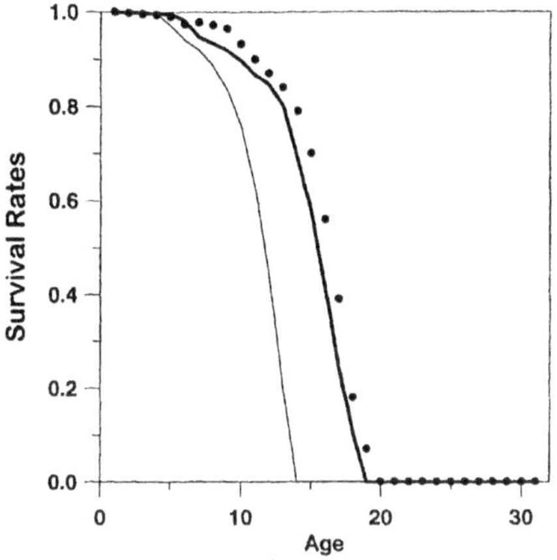

Fig.2.3.6 - Normalized survival rates for constant (thin line), linear (thick line) and parabolic (circles) fertilities; $T = 2$, $R = 4$ and $M = 1$. (Figure sent by Argollo de Menezes, Racco and Penna).

to animal and inherited by the offspring. The simulations started with *minage* randomly distributed between 1 and 32, and all the genomes free of mutations. Ito also introduced a reproduction risk, *rrisk*, in such a way that when an animal gives birth to a child at age a, both die with the probability *rrisk* which is age dependent and given by:

$$rrisk(a) = \left(1 - \frac{a}{32}\right)^2 \quad . \tag{2.3.18}$$

He studied the behaviour of the population for the cases where there are no genetic mutations in *minage*, only the ordinary bad mutations, or good and bad mutations. For all the cases the population converges to an optimal *minage* distribution which turned out to be a rather narrow peak at some age, close to what was usually assumed ($R = 8$) in Penna model simulations. Introducing mutations in the minimum reproduction age, that is, the *minage* of a child is selected randomly from parent's *minage* and *minage* ± 1, the distribution became broader. This last situation was applied for the case where good and bad mutations may exist.

Ito also assumed a constant reproduction risk (independent of age), but then found that an animal with a smaller *minage* gets more descendents than an animal with a larger one, and finally *minage* converges to the possible minimum of zero. Again we can see that Nature is not so kind: despite of the fact that ageing starts with reproduction, *it prefers a reproductive regime that starts earlier, producing many offspring, instead of longevity.* Individuals that intend to breed late in life may suffer from a heart attack, die in a car crash or because of too much beer, and so loose the opportunity to contribute to the perpetuation of the species.

The above statement was confirmed by BERNARDES (1996b), in the following way: Starting with a population $N(0) = 10^7$, he imposed a different genetic condition to 10% of the initial population. This different condition consists in a higher reproductive rate at the expense of a new threatening disease appearing after the reproduction age. In his simulations $minage = 10$ and the new disease was added up at age 12. The individuals belonging to the small fraction have $b = 0.6$, and the others, $b = 0.4$. A birth rate smaller than 1 simply means that individuals will generate one offspring with probability b. Bernardes obtained that the final population having the special bad gene (but with a higher b value) was greater than that without it. He interpreted this bad gene as a kind of antagonistic one, and observed that the final population with antagonistic genes has less harmful mutations in the childhood and young ages, but more harmful mutations accumulated at old ages. He concluded that natural selection prefers to spread a genetic code with more mutations, producing the loss of old individuals, but ensuring a higher amount of children in the population. (*Antagonistic pleiotropy* refers to genes that play an important role in keeping individuals alive at earlier ages but may become harmful at advanced ones).

BERNTSEN (1996) investigated the effects of an age-dependent mutation rate in the Penna Model, assuming that the mutation rate depends linearly on the age of the parent: $M = a/\alpha$, where α is a constant. As a result, the total population and the maximal age (age after which no individual was alive) were seen to decrease, while the mortality increased. In this same paper the effect of assuming weak babies was also investigated by assuming a lower threshold T for individuals with age less or equal to two. It produced a mortality curve qualitatively similar to that observed for humans with higher child mortality. An alternative model for child mortality (A. Strotmann, priv. comm., January 1997) based on a reinterpretetion of the time scale gives also reasonable results: bits 4 to 32 correspond to one year, but bit 3 to half a year, bit 2 to 3 months, and bit 1 to six weeks, thus taking into account that more growth and thus also more dangers of errors occur in childhood. (Note also the non-monotonic variation of the death rate with age in THOMS, DONAHUE and JAN, 1995, fig.3.)

We have been talking about experiences with *iteroparous* populations, i.e., individuals may give birth every year after a minimum age. Let us now see what was done by THOMS et al (1995) for a *semelparous* population (individuals reproduce only once, but not necessarily at the same age; a common reproduction age for all individuals is a particular case, observed in some species as salmon and mayfly, as discussed before). They considered a model where each individual may reproduce at a single age, randomly chosen from 10 to 32, and producing 0 to 3 offspring with equal probability. This age at which reproduction occurs is passed to the offspring. They observed that after 1000 years, only individuals that reproduce at age 10 were still alive, leading to a 10 year population cycle with a nine year gap when no new individuals are born. The simulations were performed for both mutation-free and random initial populations, leading to the same *final* result. They also simulated the case where each individual has a single reproduction age, randomly chosen to be 10, 11 or 12, but not transmited to the offspring. This gave a much more uniform population over time without a ten year birth cycle.

2.4 Comparison of the Penna Model and the Azbel theory

In the first chapter we saw that the human mortality q increases exponentially with age a, in agreement with the 19th century Gompertz law, that is (MUELLER, NUSBAUM and ROSE 1995):

$$q(a) \simeq \ln[N(a)/N(a+1)] \propto \exp(ba) \quad ,$$

where $b \simeq 0.1$/year. As also explained, the agreement between real mortality and the Gompertz law is obtained when adulthood is reached and children's diseases are overcome. Deaths due to car crashes of youths or those caused by environment constrictions should not be included. It means that in order to check if the asexual Penna model fits the Gompertz law, only deaths due to diseases are to be computed and those due to the Verhulst factor disregarded. An easy way to do this is to use a normalized mortality , in the same way as we used, in section 2.3.1, a normalized survival rate :

$$q(a) = \ln\left[\frac{N(a)}{N(a+1)} \div \frac{N(0)}{N(1)}\right] \quad . \tag{2.4.1}$$

Fig.2.4.1 shows the mortalities obtained with (diamonds) and without (crosses) the Verhulst factor. In Fig.2.4.2 the number of survivals as a function of age is also shown. As already mentioned in chapter one, as far as we know the

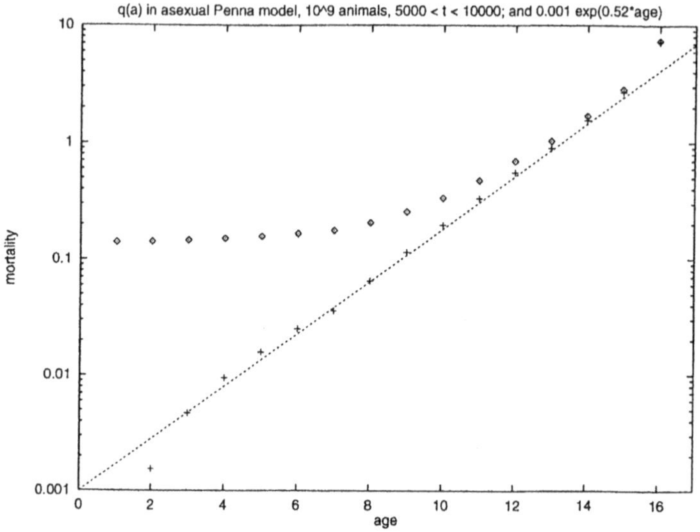

Fig.2.4.1 - Mortality as a function of age. Crosses: deaths due only to mutation accumulation; diamonds: deaths due to the Verhulst factor are also included. The dashed line corresponds to the fit $q(a) = 0.001 \exp(0.52 * a)$, in agreement with the Gompertz law. Parameters: $T = 3$, $R = 8$, $b = 1$, $M = 1$ and $N(0) \sim 10^9$ individuals. The mortalities were averaged from timestep 5000 to timestep 10000. The deviations from the Gompertz line at young age are real and not just fluctuations; but here the neglected child mortality dominates anyhow.

Penna model is the first Monte Carlo model for biological ageing that results in the Gompertz law.

The phenomenological Azbel theory , based on mortality curves of demographic data, states that

$$q(a) = Ab \exp[b(a - X)] \quad , \tag{1.3}$$

with A, b and X related by the equation:

$$c = \ln A - bX \quad . \tag{1.4c}$$

Observing that X and A are the same within one species ($A = 11 \pm 2$ and $X = 103 \pm 1$ years for humans) Azbel proposed that they must be genetic factors. Equation (1.3) is valid for sufficiently homogeneous cohorts, that is, individuals

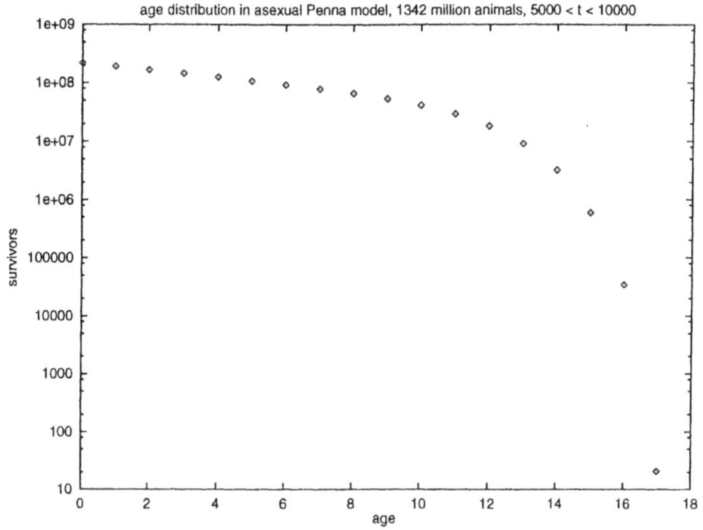

Fig.2.4.2 - Number of survivals as a function of age, for the same parameters of Fig.2.4.1.

with the same genetic factors that will affect their life history, giving smooth death rates. When the population is heterogeneous, some features may change like the appearence of arbitrarily old Methuselah's and a linear increase of reciprocal mortality with age for sufficiently old ages $a > X$ (eq.1.5a).

In order to check if the Penna model agrees with Azbel results, RACCO, ARGOLLO DE MENEZES and PENNA (1998) performed many simulations using genomes of 128 bits. Firstly, they kept fixed all the parameters of the model, changing only the random seeds, obtaining many different pairs (c, b) (counting only deaths due to mutations and for ages > 15). After that, they chose a different mutation rate M and again performed the simulations. Their results are shown in Fig.2.4.3. It can be seen that for the same value of M, all pairs (c, b) fall in the same line, and that different values of M give different slopes, suggesting that this parameter can be used to characterize different species in the Penna model.

For each straight line of Fig.2.4.3 the corresponding value of X was computed, and plotted as a function of M. The result is shown in Fig.2.4.4, and corresponds to the fitting $\log_{10} X = 1.42 - 0.21 \log_{10} M$. With this relation between M and X the authors found that in order to have $X = 103$ years, one needs $M \simeq 0.001$, in agreement with the mutation rate proposed before to fit

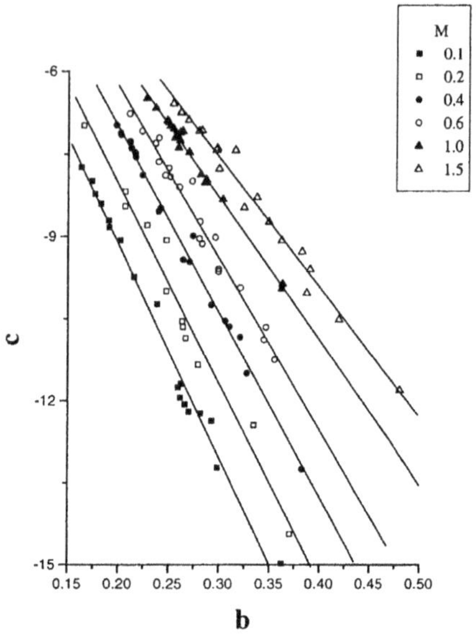

Fig.2.4.3 - Linear behaviour of the death rate parameters (c, b) obtained for different mutation rates M. The birth rate = 0.5, $R = 12$ and $T = 3$. Each point of the plots corresponds to a simulation with a different seed. (Figure sent by Racco, Argollo de Menezes and Penna).

the female mortality in Germany (PENNA and STAUFFER 1996). Analysing their results, Racco et al observed that the characteristic age X is indeed smaller than the maximum lifespan, as obtained by Azbel, and also that b increases with M.

We also performed huge simulations to confirm the agreement between the Penna model and Azbel's theory . We ran 512 simulations, each one starting with 2 million individuals, for a single value of the mutation rate $M = 1$. Our results are shown in Fig.2.4.5. It is possible to note, comparing figures 2.4.3 and 2.4.5, that for large populations the range of c and b values is smaller than that for small populations.

In the same paper quoted above, Racco et al propose that an *heterogeneous* population can be regarded as a population out of equilibrium, where very different genomes exist together (many of them will disappear later, only

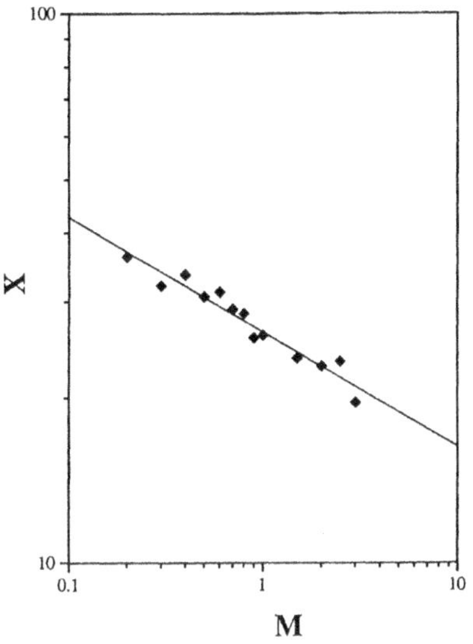

Fig.2.4.4 - Power law behaviour of X with respect to M. The solid line corresponds to the fit $\log_{10} X = 1.42 - 0.21 \log_{10} M$. Parameters are the same of Fig.2.4.3. (Figure sent by Racco, Argollo de Menezes and Penna).

remaining those with a small number of mutations before the reproduction age) . With this in mind, they measured the distribution of individuals as a function of age, for a population still far from equilibrium, and indeed found a non-exponential behaviour (equivalent to that observed for medflies) instead of the exponential one observed in Fig.2.4.2.

The idea of representing an heterogeneous population by a non-equilibrium one still needs, in our point of view, further experimental checks. However, for homogeneous stable populations it seems that the Azbel theory and the Penna model are in surprisingly good agreement for humans, reinforcing the mutation accumulation theory of biological ageing and the reliability of the Penna model.

Although it is hard work to put together all features of the asexual Penna model, we hope that at least the most important ones have been explained in

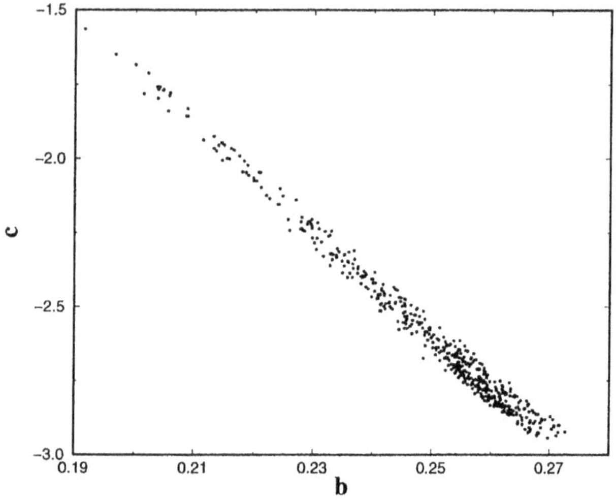

Fig.2.4.5 - Linear behaviour of the death rate parameters (c, b) for a single value of $M = 1$. Each of the 512 points corresponds to a simulation starting with 2 million individuals. The other parameters are: $T = 3$, $R = 8$ and birth rate $= 1$.

the previous sections. Let us now, and finally, talk about sex!

2.5 The Sexual Version of the Penna Model

Why does sex exist?!? This seems to be a silly question, that could be immediately answered by a teenager whose mainly interest is chasing girls: because it is good! Without it, we would die of boredom! However, not only the senile author of this book is still alive but also many species, that have been surviving for thousands of years without sex. Bacteria and virus are examples of organisms that reproduce by division, after producing a copy of its own genome (mitosis). Sex implies recombination, that is, a mixing of parental genes in progeny. The basis of sexual recombination is the production of gametes by meiosis and their subsequent union to form a zygote . A gamete is a haploid germ cell having a single set of unpaired chomosomes. All other cells in our body are diploid cells, containing pairs of chromosomes , called homologous, in its nucleus. The union of two gametes , one from the mother and the other from the father, produces a diploid zygote containing hereditary information from both parents. We can visualize the process considering that our genetic features are stored in a pair of (homologous) strings, present in all our cells, except in gametes, that contain only one string (half of the chromosomes of other cells). However a gamete does not correspond to a simple division of the two homologous strings into two independent ones. To produce a gamete there is first mitosis , a process where the two strings of the homologous pair duplicate (resulting in two pairs of homologous strings). After that, the strings cross over, breaking at identical positions. The broken ends exchange, recombining the genes, in such a way that at the end there are again two pairs of homologous strings, but now both different from the original ones. Then the meiosis (division) occurs, originating four gametes (strings), each with half the number of chromosomes . When a gamete from the father joins the one from the mother a zygote is produced containing again two homologous strings. The end result is that the zygote has part of the genome from the mother and the remaining part from the father.

Life would be easier (at least for computational physicists) if recombination could be found only in sexual reproduction . However, it is possible to find many species that also contain pairs of homologous strings, that recombine one with the other producing an offspring, different from the parent, but involving only one individual. This kind of reproduction is called *meiotic parthenogenesis* . In Nature all those mechanisms can be found, from simple asexual reproduction (copy) to the sexual one. And so now the question arises again: why sex? The most common explanation is that the recombination process may clean the genomes, giving a chance to the offspring of not having a given inherited disease present in the father genome, for instance. However, in the same way it can insert defects that were not in the parent before. And more than that, recombination may exist, exhibiting this same properties, without involving sex. There seems to exist at least one common hypothesis for sexuality: it produces a high degree of diversity, that can be very important to adaptability in a continuous fluctuating environment. There are species that reproduce either sexually or asexually: typically, rapid asexual reproduction occurs at the beginning of the growing season while sexual reproduction occurs only at the end of the season when conditions are clearly destined to change (ROUGHGARDEN 1996, STEARNS 1987).

Of course we don't have answers for all these questions; what we are going to present in the following section are some aspects of sexual reproduction that have been studied through computer simulations and that may help us to understand a little bit more about reproduction in general. Before, let us explain how sex was introduced into the Penna model. The first one to do that was BERNARDES (1996b), in the following way. The genome of each individual is represented now by two strings, *to be read in parallel*, i.e., to a given age correspond two bits, one from each string. First problem: how to compute diseases in order to compare with the maximum allowed number T? Here appears the important concept of dominant and recessive mutations. When an offspring inherits a given characteristic from both the mother and the father, it is called an homozygote ; in this case there is no doubt that the offspring will present the same common characteristic of the parents. In the Penna model it corresponds to an offspring that has two zero bits or two set bits at a given position (age). Independently of the nature (dominant or recessive) of this position, for the individual with two zero bits no disease is computed in that year, and for the individual with the two bits set a new disease appears at that age, to be added to the previous ones and the total amount compared with T. For heterozygotous individuals, corresponding in the model to those presenting different bits at the same position of each string, a new disease is computed only if that particular position is a dominant one. There are two ways to decide if a given position is dominant or not. In the first one (BERNARDES 1996b), each position has a probability h, varying in the interval (0,1) and defined a priori, of being dominant. The second way, which we are going to adopt, was introduced by STAUFFER et al (1996): in the beginning of the simulation, an extra word (also containing 32 bits and called DOMINANCE in the computer program - appendix 2) is defined, containing 1 bits only at the corresponding dominant positions. These positions are randomly chosen, but their number is defined a priori. This word is kept fixed during the whole simulation. Both strategies give similar results.

Let us now see reproduction and the offspring genome. As in the asexual case, there is a minimum reproduction age R when females start to generate b offspring every year. Of course they first choose, at random, a male to mate (since men are indeed all alike, this makes no difference); the only restriction is that his age must also be equal to or greater than R. The offspring genome is constructed in the following way: the two strings of the male, for instance, are broken in the same random position, and the complementary pieces, belonging to different strings, are joined to form two male gametes . One of the gametes is then randomly chosen to be passed to the offspring. After that, $mmut$ random mutations are introduced into this gamete, and the final result corresponds to one string of the baby genome. The same process occurs with the female genome, generating the second string of the baby, with $fmut$ mutations. At the end the offspring genome contains a total of $M = mmut + fmut$ mutations. Finally, the sex of the baby is randomly chosen, with a probability of 50% for each one.

Stauffer et al also included a pregnancy period into the model, as well as a kind of male fidelity. However, no significant difference appears in the results because of these parameters. Details can be found in appendix 2.

2.5.1 The Dominance Effect and the Redfield Model

Our interest in sexual reproduction was sparked by the paper entitled "Male mutation rates and the cost of sex for females", by REDFIELD (1994), ignored in SCIENCE (1998). In her paper she concludes that because male gametes are produced in a high amount and present a mutation rate much higher than the female ones, it could be better if females reproduce asexually and forget men (it is known that they are disgusting, but useless would be too much!). She suggested that females could choose the youngest males available (an excellent idea, according to the female author of this book), and in so doing the contribution from male mutations might be reduced and sex could be advantageous. Her algorithm is a very fast one, avoiding all random numbers and calculating from the probability distribution of mutations at time t the corresponding distribution at the next time step $t+1$. It is assumed from the beginning that the population is constant; thus the birth rate must adjust to the temporal variation of death rate. (Thus, the crucial effect of mutational meltdown , undoubtedly present in Nature, cannot be directly studied with this assumption.)

In Redfield's program the population is characterized by a distribution of mutations $P(m)$, $m = 0, 1, 2, ...$, giving the probability that an individual has m genetic diseases in the genome. Darwinian selection of the fittest then transforms this $P(m)$ into a survivor distribution $L(m) \simeq (1-s)^m P(m)$ giving the probability that a survivor has m deleterious mutations , where $s = 0.1$ is a selection coefficient. In the asexual case n new hereditary mutations happen according to the Poisson distribution $\mu^n \exp(-\mu)/n!$, where $\mu \simeq 1$ corresponds to the mutation rate per genome per generation. Now the individual has $m+n$ mutations, and the corresponding distribution of mutations is already the new progeny distribution $P(m)$.

In the sexual case, after selection has transformed $P(m)$ into $L(m)$, mutations of rate μ produce the female distribution $F(m)$ and those of rate $\alpha\mu$ give the male distribution $M(m)$. Then male gametes are produced containing half of the male genome; thus their mutation number m_m is roughly half of the number of mutations in the father's genome. Analogously, the number m_f of mutations in the female gametes is roughly half of the mother's number of mutations. The fusion of the two gametes adds these two numbers,

$$m = m_m + m_f \quad , \tag{2.5.1}$$

to produce a new progeny distribution $P(m)$. The cycle is repeated until changes in each distribution become negligible. In this model the fitness is the average survival probability

$$Fit = \frac{\sum_m P(m)(1-s)^m}{\sum_m P(m)} \quad ;$$

its steady state value is $\exp(-\mu)$ in the asexual case and $\exp(-\mu(1+\alpha)/2)$ in the sexual case. Even for $\alpha = 1$ sex does not improve at all the average fitness and thus fails to justify its additional cost and complication; $\alpha > 1$ makes sex worse.

It is obvious from eq.(2.5.1) that Redfield model does not distinguish between dominant and recessive mutations. Considering that only a small fraction h of the genetic diseases are dominant, and a fraction proportional to r are recessive, STAUFFER et al (1996) replaced eq.(2.5.1) by:

$$m = (m_m + m_f)h + r m_m m_f \quad (h \ll 1, \quad r \ll 1) \quad . \tag{2.5.2}$$

The original Redfield model has $h = 1, r = 0$, and from this point of view corresponds to the limit of extreme inbreeding, which is known to be bad. With eq.(2.5.2) the advantages of sexual compared with asexual reproduction become obvious even at the high value $\alpha = 10$ of male unreliability: for $\mu = 0.3$, asexual reproduction gave a fitness of 0.74, whereas sexual reproduction gave a fitness of 0.95 to 0.97 with $h = 0.2$ and $0 \leq r \leq 0.2$. Using eq.(2.5.1) Redfield obtained the small fitness 0.19 for this same sexual case. Now we may say that the cost of sex is more than justified when compared with the boring asexual reproduction.

In some sense we may thus compare the advantages of sex with the back-up diskettes (male) on which careful computer users store important files from their hard disk (female). If due to an error (mutation) something is wrong with the file (genome on the hard disk), the correct information is still available on the diskette. But in contrast to human intervention, which then replaces the faulty hard-disk file by the unmutated diskette file, this sex model does not have such a direct DNA repair mechanism (BERNSTEIN 1977); it merely hides the bad effects of the error if the mutation was recessive, and due to random recombination the error might later get lost somewhere.

As mentioned before, Redfield suggested instead to justify sex with the assumption that females select only the youngest males for mating (from experimental observations, this is usual for men: the older they are, the younger the women they look for). Since there is no age-structure in this model, Stauffer et al followed BERNARDES (1996a and 1997) instead, by assuming that females select only the healthiest males as mates (those with lower number of mutations). Then indeed similar drastic advantages of sex were found, even if eq.(2.5.1) is used.

For the glory of the male authors of this book, MARTINS and PENNA (1998) looked at the advantage of females betraying their husbands. In a bird species there is a tendency of the extra-pair male to be older than the social male (suggesting a process of learning). Martins and Penna simulated the behaviour of these birds, using the Penna model, in the following way: each female generates offspring from the husband, and also from an extra-pair male randomly chosen, that must be older (by some factor) than the husband. They obtained that the offspring generated by the extra-pair relation have a higher survival probability than the husband's offspring. More than that, the survival probability of the birds that were exclusively husbands is lower than that of the birds that were extra-pair males. In this way, it seems that the Redfield's idea of women selecting only the youngest males for mating should be replaced by the selection of the oldest males (despite the protests of the female author of this book). At least in the Penna model, older individuals have healthier offspring than younger parents since they must have survived to that older age.

2.5.2 Sex and Catastrophic Senescence

In sections 2.3.2 and 2.3.3 we showed that life stops after the maximum reproduction age, due to mutation accumulation in all bits beyond this age. There, we presented the question of how to explain the observed higher survival rate for women even though they go through menopause . The explanation cannot appear from a Redfield-type model since an age structure is necessary. Using the sexual version of the Penna model already explained, STAUFFER et al (1996) studied the problem considering the same minimum reproduction age R for both males and females, but different maximum reproduction ages $mage$ and $fage$, respectively. They observed that for $R = 10, fage = 12$ and $mage = 32$, both males and females live until about an age of 16, roughly the same life expectancy as for the asexual case for a maximum reproduction age = 32. Fig.2.5.1 shows some of the obtained survival rates as a function of age. The common parameters are: $R = 10$, $T = 4$, $N(0) = 100,000$ (half for each sex in case of sexual reproduction) and $N_{max} = 10N(0)$; birth rate $b = 2$ in asexual case and $b = 4$ for females in sexual case (giving $b = 2$ per individual as in the asexual one). Squares and x: asexual reproduction, maximum reproduction age = 12; squares correspond to mutation rate $M = 2$ and x to $M = 1$. Diamonds and +: sexual reproduction, $M = 2$ (one mutation from each parent) and dominance = 6/32 (which means that 6 in a total of 32 positions correspond to dominant diseases); diamonds correspond to $fage = 12$ and + to $fage = mage = 32$.

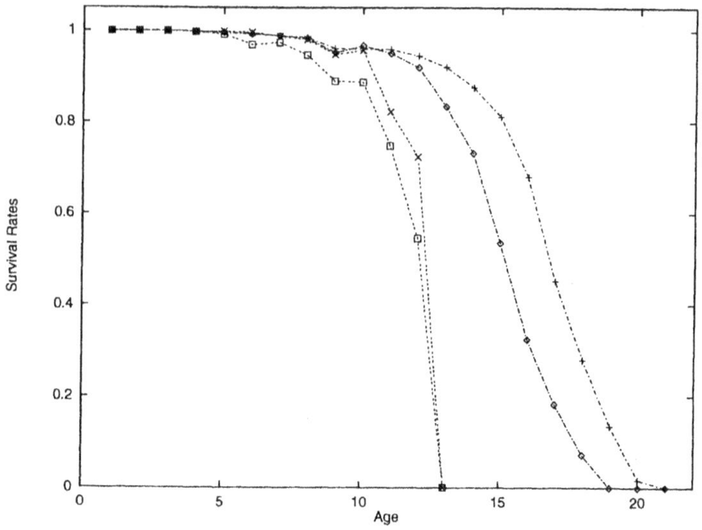

Fig.2.5.1 - Normalized survival rates for sexual and asexual reproduction. Squares ($M = 2$) and x ($M = 1$) correspond to asexual reproduction from age 10 to 12. Diamonds and + correspond to sexual reproduction; diamonds for reproduction from 10 to 12 and stars from 10 to 32.

If we compare the two sexual cases presented in the figure, we note that life expectancy is shorter if menopause is considered, but there is no catastrophic senescence: menopause sets in at age 12 and the total population survives until age 19. A crucial aspect of the model as well as in Nature is that *sex is not transmitted genetically*; independent of the genome we take each child as male with probability 1/2, and as females otherwise. So if death is hidden in the offspring genes, then either both males and females die soon, or both males and females die late. It is important to note that with this version of the model males and females present exactly the same survival rates , even if the male mutation rate is larger than the female one, and so neither Nature nor the model allows females to die sooner from accumulated genetic mutations than the males. Thus, we may conclude (for the moment) that men might be useful for something.

The reader may ask why in Fig.2.5.1 there are two asexual cases, one for $M = 1$ and the other for $M = 2$; the reason is that if we want to compare sexual with asexual reproduction , we have to choose properly the parameters of the model. If we admit in the sexual case $M = 2$, one mutation coming from the mother and the other from the father, which is the correct value of M, in the asexual case, to make a comparison? One possibility is simply to assume $M = 2$ in both cases. However, it is also possible to consider that a higher mutation rate is the price to be paid for sexual reproduction , that involves two strings instead of only one. From Fig.2.5.1 we see that it makes no difference for the catastrophic senescence effect. In Fig.2.5.2 we present some comparisons for the general case, where the maximum reproduction age is 32 for both sexes, that makes more clear the differences (or comparisons) between sexual and asexual reproduction. In this figure the general parameters are: $R = 10$, $fage = mage = 32$, $b = 2$ and $T = 4$. Squares and +: asexual reproduction, $M = 1$ (+) and $M = 2$ (squares); diamonds and line: sexual reproduction, $M = 2$ (one from each parent), dominance = 6/32 (diamonds) and dominance = 32/32 (line, corresponding to the case where all diseases are dominant). From this figure we see that even considering that a double number of mutations is the price for sexual reproduction , sex is advantageous if recessive mutations are considered.

2.5.3 Survival Rates in Real Populations

As mentioned before, it is known that women have a slightly smaller death rate than men (see Fig.1.1) until around 90 years, when both come together. However, this feature does not appear in the traditional sexual Penna Model. The first attempt to obtain such effect was made by MOSS DE OLIVEIRA, DE OLIVEIRA and STAUFFER (1996), through the inclusion of somatic mutations in the model, which are not transmitted to the offspring. (Increasing the *genetic* mutation rate for males but not for females *does not* produce a difference *between* the two death rates, again because sex is not transmitted genetically). Somatic mutations can happen anywhere in the body, and thus occur with much higher frequency; on the other side they are less effective. Thus it is possible to neglect fluctuations in the number of somatic mutations and assume e.g. that each year the survival probability is reduced by a given percentage due to somatic mutations. Again excluding deaths caused by the Verhulst factor (eq.2.4.1), the result is shown in Fig.2.5.3. In this figure somatic mutations reduce the lifespan by the same amount as an hereditary

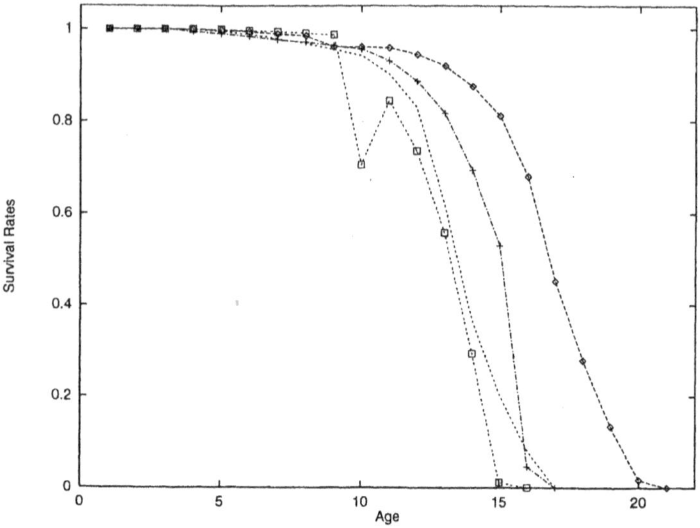

Fig.2.5.2 - Normalized survival rates for sexual and asexual reproduction. Squares and + refer to asexual reproduction for $M = 2$ and $M = 1$, respectively; diamonds and line refer to sexual reproduction, for dominance equal 6/32 and 32/32, respectively.

mutation but only with probability of 0.65 for males (squares) and 0.45 for females (circles). It can be seen that only at old ages, where the deaths due to hereditary mutations dominate over those from somatic effects, do the death rates of males and females approach each other, just as in real world of Fig.1.1. Using this same strategy, but adopting very low mutation rates PENNA and STAUFFER (1996) obtained death rates comparable to those found in Germany. They assumed hereditary mutation rates of 0.01 and 0.002 for males and females, and somatic mutations diminishing the survival probabilities each year by 0.01 and 0.002 percent for males and females, respectively, and $T = 2$ for both sexes.

We can also see from Fig.2.5.3 that the Gompertz law behaviour for sexual populations is not so well reproduced if compared to asexual populations (Fig.2.4.1). Using an initial population of $2 * 10^8$ individuals, we obtained a better result, as shown in Fig.2.5.4. The death rates as a function of age were averaged from $t = 10^4$ to $t = 2 * 10^4$, and somatic mutations were not considered, giving equal survival rates for both sexes. Anyway one sees an S-shape curve that slightly deviates from a straight line.

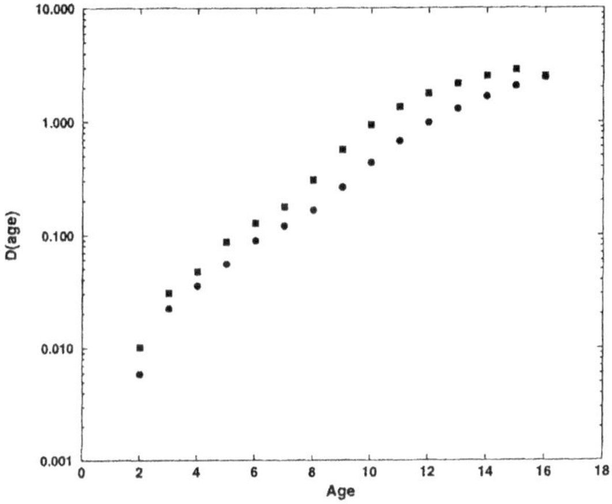

Fig.2.5.3 - Mortality as a function of age, disregarding deaths due to the Verhulst factor. Somatic mutations are included every year, with probability 0.65 for males and 0.45 for females. Including the somatic mutations, the female mortality (lower data) is lower than the male (upper data) one, both curves coming together only for advanced ages. Parameters: female reproduction from age 6 to age 15; male reproduction from 6 to 32; mutation rate 2 for males and 1 for females (per string per generation); dominance = 6/32 and $T = 3$.

Another approach to obtain different survival rates for males and females was used by PENNA and WOLF (1997). They assumed that this difference can be genetic, and adopted different thresholds for each sex. Using $T = 3$ for males and $T = 4$ for females, they obtained the observed behaviour of a higher death rate for men, both curves reaching the same value at advanced ages. The biological motivation was the fact that for some kinds of cancer, as lung cancer for instance, there are 95,400 dead males for 62,000 females (from National Cancer Institute statistics, USA (1995), http//www.nic.nih.gov). Due to the fact that there are some diseases that affect only one sex, like breast cancer (46,000 cases) for females and prostate cancer (244,000 cases) for males, they also tried a different strategy choosing different dominances, the males having more bits set than the females. In fact, different dominances were already adopted by BERNARDES (1996b), and again the same expected behaviour

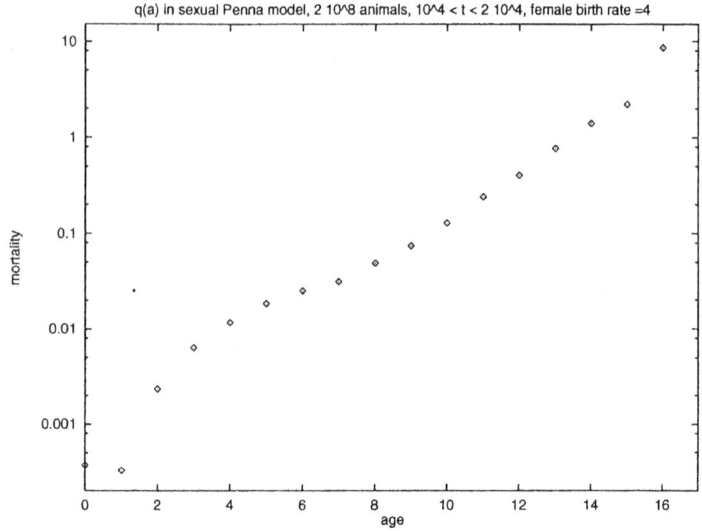

Fig.2.5.4 - Mortality versus age for the sexual Penna model, starting with $2 * 10^8$ individuals. The S-shape curve slightly deviates from a straight line, if compared to the asexual case of Fig.2.4.1. Parameters: $T = 3$, $R = 8$, female birth rate $b = 4$ and $M = 1$.

was obtained.

2.5.4 Ancestors and the Eve Effect

Although not mentioned before, BERNARDES and STAUFFER (1995) showed that for long enough times all survival individuals of an *asexual* population will be offspring of a single ancestor. They called family the group of individuals having a common ancestor. They started with N individuals representing N families; after some time the number of individuals stabilized whereas *the number of families decreased inversely proportional to* the number of iterations, *the time*. DE OLIVEIRA, MOSS DE OLIVEIRA and STAUFFER (1997) studied this same "Eve" effect also in *sexual* populations, and using different versions of the Penna model. For the sexual case we sticked a label (family name) in every female, at the beginning of the simulation, and this label was transmitted to the female offspring. We also obtained, as for the asexual case, that after many generations all survivors have the same label. Moreover, the number of families decays with time as a power law $1/t^z$, with a dynamical exponent

z roughly equal to one. In Fig.2.5.5 we show the result ($z = 1.16$) for the case where both males and females can reproduce from age 10 until death. Assuming that females stop to reproduce at age 12 (menopause), we got the same exponent. Considering the same maximum reproduction age 12 for both sexes, we got $z = 1.06$.

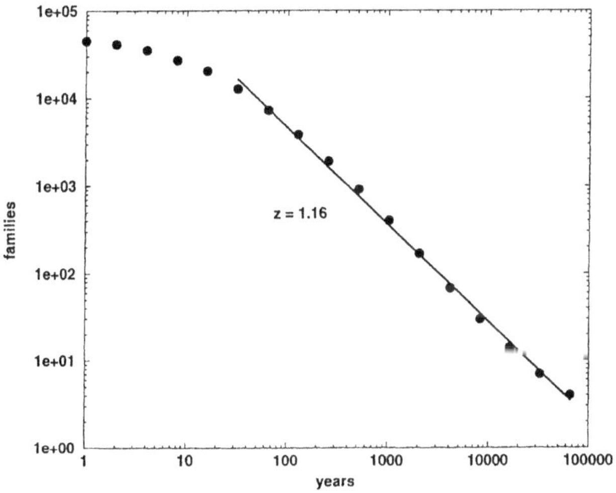

Fig.2.5.5 - Number of families (ancestors) as a function of time for the sexual Penna model. Parameters: reproduction from 10 to 32 for both sexes, $b = 4$, $M = 2$ for both sexes and dominance = 4/32.

This dictatorship of one Eve is also computationally relevant, if we use a parallel computer to simulate a single huge population, distributed among the many processors. After a sufficiently long equilibration time the one processor on which Eve lived originally, will carry the whole population, while the others are idle. Thus load balancing is needed, shifting survivors from overloaded processors to less busy ones. This was achieved by MEISGEN (1997). But even for huge populations one does not get a smooth mortality curve since Eve's offspring all had some bits in their youth set randomly, and these bits are then set for all survivors. Thus Meisgen's data are not entirely smooth. The smoother data of Fig.2.4.1 were obtained instead by averaging over 512 separate populations of smaller size than Meisgen's total population, on 512 Cray-T3E processors.

It is very important to note that *the number of families has nothing to do with the diversity of the genomes.* We measured, for sexual reproduction , the mean correlation between the common ancestor genome and the descendent ones as a function of time, and found that it goes to zero much before the number of families go to 1. That is, despite of descending from a common ancestor, after many generations there is no longer a correlation between the ancestor genome and those of the current population. We defined the mean correlation as the average overlap between the common ancestor genome and its descendent ones minus the average overlap between the common ancestor genome and the non descendent ones.

Since both sexual and asexual cases give the same Eve effect, it is reasonable to ask if it can not be a more general effect, not restricted to the Penna model parameters. In order to check this possibility, we adopted two simplified versions of the model, firstly removing the age-structure and then, also the mutations. To remove the age-structure we considered a Redfield-type model where the genomes are still represented by bit-strings , but the survival probabilities are given by $P_s = (1-s)^m$, withs $s = 0.1$ and m equal to the number of 1 bits in the genome (independent of their position). Reproduction and deleterious mutations were introduced as before, as well as the Verhulst factor. For sexual populations we obtained $z = 0.97$ for male and female mutation rates (0.2, 0.1), and $z = 0.93$ for mutation rates (0.02, 0.01) respectively. For the asexual case we got similar results, with $z = 1.08$ for a higher mutation rate and $z = 0.96$ for a smaller one.

Finally, we considered a model where only reproduction and deaths from Verhulst are taken into account: each individual generates b offspring every year; each individual is characterized by a family label which is transmitted to the offspring and all individuals have the same probability to stay alive dictated by the Verhulst factor. Again we got the same result, with $z = 0.98$. In this way we may say that generally if there is reproduction together with some kind of selection, after many generations one always remains with one common ancestor. Such a result holds for a general class of processes called "branching processes", as already pointed out by Kolmogorov in 1938 (see for instance, HARRIS 1989 and also ZHANG, SERVA and POLIKARPOV 1990).

More recently it has been obtained (MOSS DE OLIVEIRA et al 1998) that considering also a paternal name in sexual populations of the Penna model, that is, each offspring inherits a paternal name from the father and a maternal one from the mother, the same effect is observed. After many generations all individuals are descendents from common Eve-Adam ancestors. The power-law behaviour gives now an exponent z roughly equal to two. They have also measured the number of families of a given size N_s as a function of the size s (that is, as a function of the number of individuals the family has had during its whole existence). They have obtained a power-law distribution $N_s \propto 1/s^y$, with $y \sim 0.5$, considering sexual reproduction and that maternal family names are transmitted to female offspring (Fig.2.5.6). Such a result can be interpreted as a self-organized criticality of evolution to the edge of chaos , and has been obtained also for the simpler case where only the Verhulst factor and reproduction are taken into account, with the same exponent 0.5.

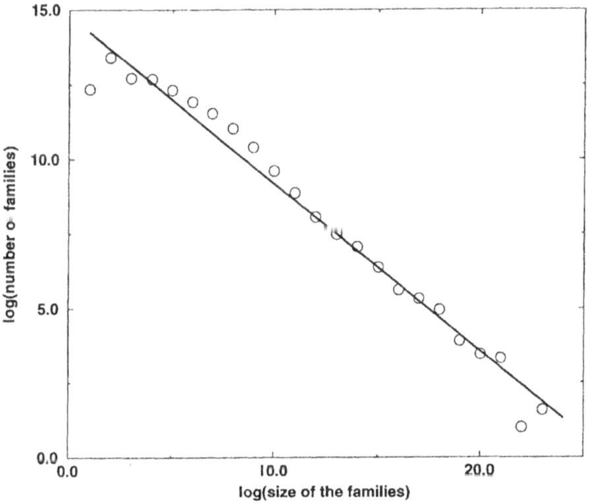

Fig.2.5.6 - Number of families of a given size as a function of the size, in a logarithmic scale (base 2), for sexual reproduction, considering maternal family names. Parameters: $N(t=0) = 200,000$, $T = 4$, $M = 4$ (two mutations per string), $b = 4$ and $R = 10$.

The "Eve effect" had already been speculated by biologists some years before, when they even suggested that Eve could have lived around 200,000 years ago in Africa (for details see OZAWA 1997 review or EXCOFFIER 1997; for malaria see RICH et al. 1998). These conjectures appeared because they searched among the current existing *mitochondrial* DNA for common sequencies. This mitochondrial DNA is genetically transmitted by females, and does not present recombination mechanism. In this way, germline mutation seems to go down to posterity as maternal inheritance from our common ancestor. It also has a very short size if compared with the DNA we have in our cell nuclei. However, there has been a controversy about this proposal: because of the limitation of restriction enzyme assay, the technique used in the majority of these studies, the information derived is restricted to only $\sim 9\%$ of the total mitochondrial DNA sequence. Hence, the standard base sequence of the common ancestor has remained unknown. On the other side, researches using PCR (Polymerase Chain Reaction) seem to confirm that indeed we come from a common ancestor that lived in Africa around 150,000 years ago (WALLACE 1994). This is a typical situation in which computer simulations may help. We can interpret our family name as the mitochondrial DNA with zero mutation rate. In order to check for how long time correlations between an original DNA and those of the current female population may persist, one has to introduce mutations in the family names, and represent them not by a label but by a bit-string of a given size. Simulations in this direction will be done by the authors, if they are still alive when this book is finished.

2.5.5 Other Types of Reproduction

Although human populations do not present sexual cannibalism (at least until now), many other species do, like some scorpions and arthropods. Sexual cannibalism is the behaviour in which the female kills the male just after (or sometimes during) copulation (one of the authors think that this is just what some men deserve; the other two see it is a constant nightmare). VANDEWALLE (1997) studied this kind of reproduction using the sexual version of the Penna model, with some modifications. He considered only dominant mutations, that is, only those presented in both strings (at the same position), and introduced a survival probability given by $f = (1 - s)^m$; m is the number of dominant mutations and $s = 0.1$ is the selection coefficient. Cannibalism occurs after the whole mating-reproduction procedure. Each male having participated in the reproduction step is killed with a probability P_{can}, considered equal to 1. Firstly, Vandewalle observed that without sexual cannibalism, the population grows exponentially towards a maximum value, which is close to the carrying capacity N_{max} and depends on the birth rate b. Then, the population decreases to a stable population size N_0 which also depends on b. For different sizes N_{max} of the ecosystem, the ratio N_0/N_{max} is independent of N_{max}. The stable normalized size N_0/N_{max} of the population increases with b and saturates to a value close to 0.77 ± 0.01. According to Vandewalle,

N_0/N_{max} seems to reach this saturation value following a power law given by

$$(0.77 - \frac{N_0}{N_{max}}) \sim b^{-3/2} \quad .$$

When the sexual cannibalism is considered, the stable population seems to saturate to a value close to $N_0/N_{max} \approx 0.69$, and this saturation seems to be reached also by a power of b with an exponent 3/2. In this case, if the birth rate $b \leq 2$, the population vanishes. However, for $b \geq 2$, the population becomes stable as for the non-canibalism case. The balance male/female reaches a value different from unity (and independent of b), since females live longer than males. This value depends on the mutation rate, and Vandewalle obtained that for a stable population :

$$f_0 + \frac{1}{\alpha} = 1; \qquad \alpha = \frac{N_f}{N_m} \quad ,$$

where f_0 is the baby survival probability (defined above). A quite surprising result is that the female population presents a peak (explosion) before stabilizing, when females cannibalize the males, although the average survival probability presents exactly the same decrease (independent of b) as for the non-cannibalism case.

Returning now to less dangerous subjects, we have mentioned in the beginning of section 2.5 that it is possible to find in Nature species that also contain pairs of homologous string, but reproduce asexually, which is called *meiotic parthenogenesis* . The offspring genome results from a crossing and recombination of the single parent genome, in such a way that the genetic features of the offspring are different from those of the parent. However, depending on the way the offspring genome is constructed after the crossing, the final result can be similar or not to that of pure asexual reproduction . For instance, DASGUPTA (1997) compared two reproduction systems, both considering two bit-strings per individual. In one case, in order to produce the offspring there is a crossing in a random position of the two bit-strings of the parent, and one piece of the first string is joined to the complementary piece of the second string. This produces one of the offspring bit-strings. The other one is produced by joining the remaining two complementary pieces of the parent genome. After that, each string suffers M randomly deleterious mutation. This strategy is similar to the one adopted for sexual reproduction , but involves only one parent. In the second case Dasgupta assumed the same strategy, but without crossing (which is equivalent to pure asexual reproduction, with no recombination). Comparing the survival of both populations after a natural desaster has been promoted, he got similar behaviours in both cases.

The natural desaster was simulated by setting the fifth-bit of both bit-strings of each individual. The total populations were found to be nearly the same, before and after the desaster.

If we think carefuly will find that this kind of recombination between strings of a single individual is not able to "clean" the genomes, differently from real sexual reproduction. Let us assume, for instance, that the parent genome is given by 001010 and 100110, and that positions 1 and 3 are dominant . Such a parent would have a total of 3 diseases. Suppose now that we cross the genome between positions 3 and 4. The offspring genome will be 001110 and 100010, containing also 3 diseases, even without introducing new deleterious mutations . In this way, this recombination may clean one string, but the bill will be paid by the other string which becomes dirty. In sexual reproduction, however, only one of these two strings is passed to the offspring: on average, one offspring is better, while the other is worse. Natural selection takes care of perpetuating the lineage of the former, and not of the latter.

A different type of recombination was considered by BERNARDES (1997), following the explanation from STEARNS (1987) that the meiotic parthenogenesis produces, in general, individuals that are homozygotous in all positions (loci). Bernardes assumed the same crossing explained above, but instead of using the two bit-strings (gametes) that appear after the crossing and recombination to produce the offspring genome he choses randomly one of the bit-strings and copies it to produce the second string of the offspring. In this way the offspring has initially two equal bit-strings, different from those of the parent. After that, mutations are introduced in each string. Because of these mutations, at the end of the process the offspring genome consists of two bit-strings, slightly different one from the other. It is easy to verify that this approach may lead to a cleaning of the baby genome, if compared with the parent one. And now things become complicated, because Bernardes didn't find, for large populations, any advantage of sex in relation to this last recombination process, *except if male competition is included* (and he considered dominant and recessive mutations in both cases). He noticed, however, that the meiotic parthenogenesis population shows a higher concentration of harmful mutations in the younger portion of the genome. SA MARTINS and MOSS de OLIVEIRA (1998) studied this problem, checking which kind of reproduction regime gives a greater genetic diversity, that can be measured by the Hamming distance (number of bits which are different in a bit-by-bit comparison between the genomes). This greater diversity helps to survives desastrous changes in the environment, as shown in their simulations after uncussessful tests of DASGUPTA (1997). If no strong evidence would have appeared to support sexual reproduction , the former question of "why sex" would come again, and maybe men should start to fight against each other in order to justify their existence (or find a way to get pregnant, following Arnold Schwarzenegger).

In fact, for human beings it seems to be easier to justify sex, because it is already shown that women live longer then men, despite their menopause , thanks to the fact that men can reproduce until death and sex is not genetically transmitted. However, many problems arise when such an argument is studied more carefully. Firstly, it suggests that men should betray their wives in order for women to live longer (which makes 2/3 of the authors of this book very happy). If all men would be faithful to their wives, then after the menopause neither the husbands nor their wives could get more children, and both would die like the Pacific Salmon. This remark was pointed to us by Stanislaw Cebrat (private communication) and, in fact, in our simulations there is no long-term fidelidity, as can be noticed by the Fortran program presented in the appendix. Due to social rules, it is not very easy to sustain such a theory among humans. On the other side, according to AUSTAD (1994), many mammals show menopause when living long enough in a zoo or laboratory, *but die before menopause* in free nature.

In this way we could follow Bernarde's title and ask: can males contribute to the improvement of species? However, according to PACKER, TARTAR and COLLINS (1998) baboons and lions present menopause in free nature, and also MARSH and KASUYA (1984) give short-finned pilot whales as an example. The question which is still a matter of controverse is why does menopause exist. A possible explanation, already pointed many years ago by WILLIAMS (1957), is that due to a reproduction risk that increases with advancing ages, it is more advantageous to the females to cease reproduction in order to take care of the already born young offspring. MOSS DE OLIVEIRA, BERNARDES and SA MARTINS (1999) simulated this idea by introducing into the Penna model the following restrictions: a) there is a risk due to reproduction which is proportional to the number of current accumulated mutations (which means proportional to age, since in the model bad mutations accumulate at advancing ages); b) there is a period of parental care: offspring whose mothers die withing this period, are killed; c) the menopause age is no longer imposed, but transmitted to the female offspring with mutations. That is, all the females start with a maximum reproduction age (menopause age) equal to 32. When a daughter is born, it inherits the mother's menopause age with probability 25%, or the mother's menopause age ± 1 with probability 75%. The minimum reproduction age is still imposed and the same for both sexes, and the male maximum reproduction age is fixed at 32. With this strategy they obtained a self-organized distribution of menopause ages, as well as a period of post menopause survival. The distribution of menopause ages is shown in Fig.2.5.7. It is important to note in this figure that despite of the risk for later reproduction, there is no self-organization of the menopause ages if there is no parental care. Both ingredients are necessary to obtain such an effect.

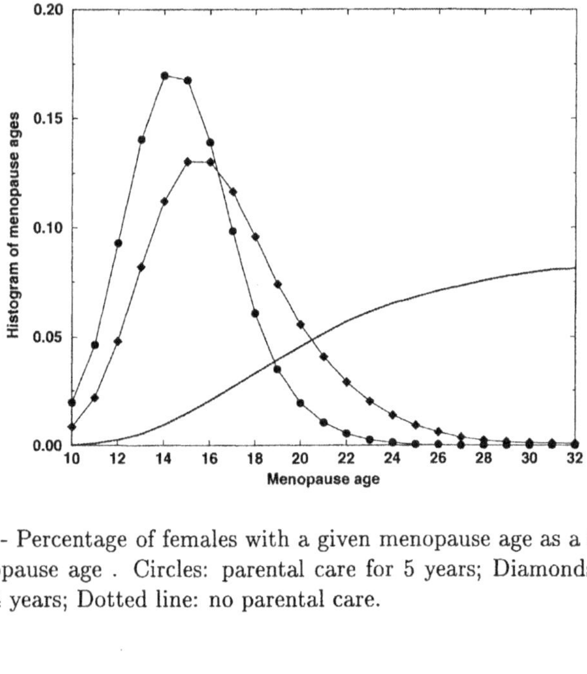

Fig.2.5.7 - Percentage of females with a given menopause age as a function of the menopause age . Circles: parental care for 5 years; Diamonds: parental care for 4 years; Dotted line: no parental care.

2.6 Analytical Approaches

In this chapter we have been concentrating in what has been done by physicists, using Monte Carlo simulations, on ageing. However, theoretical approaches have also been done, and are known since a long time. More recently, some of these theoretical approaches have been applied to the Penna model, giving results that are in a very good agreement with computer simulations. The exact results already presented in section 2.3.c are an example, but refer to a very restricted case where there is only one reproduction age and only one serious inherited disease is enough to kill the individual. The general case, considering a whole range of reproduction ages, allowing an arbitrary number of inherited diseases and the possibility of good and bad mutations has been studied by ITO (1996b), TOLEDO PIZA (1997) and DE ALMEIDA, MOSS DE OLIVEIRA and PENNA (1998), see also DE ALMEIDA and MOUKARZEL (1998). In these approaches there is a matrix responsible to evolve the population from one time step to the next one. The way the population is characterized may differ from one to the other. Unfortunately none of these calculations is simple enough to be explained in a few pages. However the general matrix idea is the same as introduced by P. H. Leslie, many years ago (see for instance CHARLESWORTH 1994 or ROUGHGARDEN 1996).

In his formulation Leslie considered four basic concepts to describe a population: age structure, population growth, age-specific mortality and age-specific fertility. Time is a discrete variable, related to periodical census of the population. If for instance, we census every ten years, then an age class must be 10 years wide. Individuals with age varying from 0 to 9 belong to the youngest class; those with age between 10 and 19 belong to the next class, and so forth. Adopting the same nomenclature used in Rougharden's book, let us call x the age class label, running from 0 to w. That is, the youngest age class is labeled 0 and the last one w. Denoting $n_{x,t}$ the number of individuals in age class x at time t, and P_x the fraction of those alive in age class x who survive from one census at time t to the other at time $t+1$, we may write the following equation:

$$n_{x+1,t+1} = P_x n_{x,t} \quad (x = 0, 1, ...w-1) \quad . \tag{2.6.1}$$

This expression allows us to predict the number of individuals in every age class, except the youngest, at $t+1$. To get the youngest age class we must use some fertility data. Let us consider F_x the average number of offspring (produced by individuals of age x) that survive to the upcoming census. If it is possible to know F_x, then the number of individuals in the youngest age class at $t+1$ is given by the sum:

$$n_{0,t+1} = F_0 n_{0,t} + F_1 n_{1,t} + ... + F_w n_{w,t} \quad . \tag{2.6.2}$$

Given the mortality and the fertility data, equations (2.6.1) and (2.6.2) can be iterated from any initial condition, giving the development of the age structure in time.

The important aspect of these equations is that *if they are iterated for a long time, a stable age distribution is obtained.* That is, the number of individuals in class x at $t+1$ is simply λ times their number at time t:

$$n_{x,t+1} \approx \lambda n_{x,t} \quad , \tag{2.6.3}$$

where λ does not depend on x and is the same for all classes. If each class increases by the same factor with every iteration, then the same happens to the total population:

$$N_{t+1} \approx \lambda N_t \quad \Rightarrow \quad N_t = \lambda^t N_0 \quad . \tag{2.6.4}$$

Since there is no Verhulst factor in this approach, we may expect an exponential growth of the population and identify λ:

$$N_t = e^{rt} N_0 \quad \Rightarrow \quad \lambda = e^r \quad . \tag{2.6.5}$$

The growth factor λ does not depend on the initial condition and is determined by P_x and F_x.

Equations (2.6.1) and (2.6.2) can be written in the matrix form

$$\vec{n}_{t+1} = L\vec{n}_t \quad , \tag{2.6.6}$$

where L is the Leslie matrix. It can be shown that λ is the biggest eigenvalue of L and that the stable age distribution is the eigenvector corresponding to λ. However, the Leslie matrix implies that one must know the F_x and P_x data. The big goal of the recent papers, mentioned at the beginning of this section, is that they give an expression for the evolution matrix without any previous knowledge about the population! The matrices are constructed based on all the possible features of the population. In the paper of de Almeida et al, for instance, a population with $N(t)$ individuals is considered, living in an environment with finite resources. The individuals present diverse genetic charges which determine different limit ages m, beyond which the individuals cannot survive. That is, m corresponds to a programed death age. The evolution matrix is, in this case, a birth matrix $A_{mm'}$, that gives the probability of an m'-parent having an m-offspring. When only one reproduction age is considered, the model predicts the catastrophic senescence effect. In fact, all the results obtained through simulations of the *standard asexual* Penna model are recovered by this analytical approach.

2.7 Summary

We have presented in this chapter the asexual and sexual versions of the standard Penna model for biological ageing, as well as the many modifications that have been introduced into the model in order to simulate specific features found in real populations. References are included in the body of the chapter. In this section we list the main features of the model, including those obtained through the modifications mentioned above.

• The Penna model is based on mutation accumulation theory; each individual presents a genome , represented by a bit-string of size B, each bit corresponding to a year in the individual lifetime. The presence of a 1-bit in a given position (year) means that the individual will suffer the effects of an inherited deleterious mutation (disease) in that and all following years. There is a dispute for space and food given by a Verhulst factor, that randomly kills individuals independently of their ages or genetic charges. It is of great importance in computer simulations, in order to prevent an exponential growth of the population. The individual dies if it reaches age B, or if the current number of diseases is greater than or equal to a threshold T, or if it loses the dispute for the environmental resources. There is a minimum reproduction age R such that, at and above it, the individual generates b offspring every year (or generates one offspring with probability b, for b values smaller than one). The offspring inherits the same genome of the parent, except for M mutations acquired at birth (section 2.1).

• The mutations acquired by the offspring can be good or deleterious. Since in Nature good mutations are very rare, and are generally related to the appearence of a new species, in general only bad mutations are taken into account. This means that the offspring genome is equal (when no mutation occurs) or worse than the parent genome. Even so, mutational meltdown (unavoidable death of the whole population due to mutation accumulation) is avoided by those offspring whose genomes are equal to the parents genomes. The inclusion of good mutations does not change the features obtained through the model. It is also possible to use an integer version of the model, in which the single bit 0 or 1 is replaced by an integer, representing several life-threatening inherited diseases . No significant difference is obtained with this strategy (section 2.2).

• The simulations may start with all the genomes free of mutations (only zero bits), or with randomly distributed genomes. In both cases, after some generations, a stable population is obtained independent of the initial condition. A stable population means that the number of individuals at any given age is constant in time. When it happens, the weaker individuals, having a great number of 1 bits before the reproduction age, have already disappeared from the population. In fact, this is the main feature of the model: only those individuals that are able to reach the reproduction age can generate offspring and so keep their lineage in the population. In order to reach the reproduction age, the individual must have its genome as clean as possible until that age. Since mutations do exist, they accumulate at old ages, generating the ageing effect. In this way, ageing means a decrease of the survival probability with age (section 2.3.1).

• An important quantity to study ageing is the age-dependent survival rate . It is defined as the average number of individuals of age $a + 1$ divided by the average number of individuals of age a. Because of the reasons explained in the above paragraph, the survival rate obtained through the model starts to decay as soon as reproduction starts. In the asexual case, if a last reproduction age is included, after many generations there are no individuals alive older than this last age. The catastrophic senescence, observed in some species like salmon and mayfly, is a consequence of this effect: the animals reproduce only once, always at the same age. For ages above this last reproduction age, there is no more selection pressure against mutations, since the individuals will not reproduce anyway. If they remain in the population, they will only dispute with the younger ones for food and space. These results can be seen in Fig.2.3.2, section 2.3.b, and naturally appear from the dynamics of the model (sections 2.3.2 and 2.3.3).

• Although senescence starts with reproduction, implying that those individuals that reproduce later also age later, nature prefers that the whole population reproduces as earlier as possible. This occurs because according to Darwinian selection only the perpetuation of the species is important, and not what may be better to particular individuals. The earlier reproduction starts, the greater is the number of offspring that can be generated (but the smaller is the population lifetime - sections 2.3.2, 2.3.4 and 2.3.6).

• It is possible to introduce privileged conditions into the model. In order to obtain the oldest old effect, for instance, some particular genomes are considered as special ones. Individuals presenting such a genome generate offspring with *good* and bad mutations. Parental care has also been introduced as a privileged condition: individuals whose parents are still alive may more esasily escape from the Verhulst factor (section 2.3.4).

• The standard Verhulst factor is given by $V = 1 - N(t)/N_{max}$, where N_{max} is the maximum population size the environment can support. It gives the survival probability of any individual according to the current population size $N(t)$. At every time-step of the simulation a random number between zero and one is generated, for each individual, and compared with the current V value. If the random number is greater than V, the individual dies. Changes in this Verhulst factor have been introduced to simulate prey and predators, the social behaviour of Alaska wolves and to study the population evolution in a periodically changing environment (section 2.3.5).

• Different strategies of reproduction have been introduced to simulate some characteristics found in particular species. For instance, a fertility rate that increases linearly with age has been used to reproduce the high maximum lifespan observed in trees. A self-organized reproductive regime has also been studied, considering a different minimum reproduction age for each individual, that is transmitted to the offspring. In this problem a reproduction risk is also included. An age-dependent mutation rate has also been investigated, as well as the effect of assuming weak babies (assuming a lower T value for ages less or equal to two - section 2.3.6).

• It can be shown that the Penna model fits the Gompertz law, which predicts an exponential increase of the mortality with age. It can also been shown that the results obtained through the model are in a good agreement with those obtained by the phenomenological Azbel theory , developed through the anallysis of real population data, and that predicts an unitary mortality law for the species. As far as we know, the Penna model is the first Monte Carlo model for biological ageing that reproduce such behaviour (sections 1.1 and 2.4).

• In the sexual version of the model the genomes are represented by two bit-strings , one coming from the mother and the other from the father. The offspring genome is constructed after crossing and recombination of the parents genomes. In order to compute the number of serious inherited diseases , and compare the result with the threshold T, dominant and recessive mutations are considered. The important role played by recessive mutations is used to compare asexual with sexual reproduction (sections 2.5 and 2.5.1).

• It can be shown that because sex is not genetically transmitted, it can avoid premature senescence of women despite the menopause they go through - thanks to men who are, in principle, able to reproduce until death (section 2.5.2).

• If the rate of somatic mutations or the number of dominant diseases is assumed to be higher for males than females, a higher female survival rate is obtained, as already observed in real populations. If one simply assumes that the genetic mutation rate M is higher for males than for females, such an expected behaviour does not appear, i.e., one gets the same survival rates for both sexes (section 2.5.3).

• Both asexual and sexual versions of the Penna model give that we all descend from common ancestors (the "Eve" or "Eve-Adam" effect). Attributing family names to each individual at the beginning of the simulation, this name being transmitted to the offspring, one always end up with a single family name, the number of lineages decreasing in time as a power law, with an exponent always close to 1 (asexual case) or 2 (sexual case). However, it has been shown that such a phenomenon has nothing to do with the diversity of the genomes. The average correlation between the ancestor genome and those of the current population reaches a constant value, much before the number of families equals one. This can be of great practical importance for simulations in parallel computers (section 2.5.4).

• The parthenogenetic reproductive regime, in which individuals reproduce asexually but contain a pair of homologous chromosomes in their genomes is considered, and compared to pure asexual and sexual reproductions. Because this kind of reproduction also presents crossing and recombination, the advantages of sex are not so easily observed (section 2.5.5).

• Analytical calculations of the original Penna model have been recently developed, in good agreement with simulation results. A brief explanation of the Leslie matrix approach and references can be found in section 2.6.

As emphasized before, the Penna model for biological ageing is extremly suitable for computer simulations and reproduces many of the features found in Nature, as the catastrophic senescence of pacific salmon and a higher survival rate for women than men, the two curves coming together above advanced ages. We do not claim that this is the only approach to study ageing, and that other theories are to be disregarded. In fact, many of the current theories are not mutually exclusive. The Penna model is the approach we have adopted to study ageing, due to it's computational advantage, and that presents some fundamental characteristics: it results in the Gompertz law of exponential increase of mortality with age, and is in a good agreement with the phenomenological Azbel theory , based on real data.

However, we know that the model has some assumptions that may not correspond to reality. Some of these assumptions have been pointed out by the geneticist CEBRAT (1997), and the most important ones are summarized below:

1) The genes that are taken into consideration by the model could be called *death genes*, and correspond to those genes that are switched on only *at specific ages.* In this way, the model does not predict the genetic death during the first period of embryo development. In other words, there are genes called *housekeeping genes* by the geneticists, that are necessary to run all indispensable functions of each living cell, and that are not being considered in the Penna model. A damage (mutation) in such a kind of gene may happen *at any moment of life.* If it happens in a somatic cell - it kills this cell. If it happens in a germ cell - it can kill an offspring even before birth.

2) In section 2.5.3 somatic mutations have been used in order to obtain a higher survival rate for females than for males. It has been assumed that each somatic mutation is equal in the effect to a heritable mutation in the same locus. Heritable mutations affect genes located in the genomes of gametes,

and somatic mutations only affect genes in the genome of somatic cells. A heritable mutation in the germ cell is transferred to the embryo and can affect *each cell* of this organism, but a somatic mutation will in principle affect only the somatic cell whose genome has been affected (although sometimes only one mutation in one somatic cell can kill the whole organism). CEBRAT (1997) proposes, to avoid this delicate problem of evaluating the rate and weight of somatic mutations, to analyse the influence of X chromosome in male and female genomes. Women possess two copies of the X chromosome, but men only one. It means that any mutation in the male X chromosome must be considered a dominant one, since it can not be balanced by an allele in the homologous chromosome. SCHNEIDER et al (1998) confirmed this prediction by computer simulations giving a male mortality roughly twice the female mortality except at old age.

Finally, we want to say that we are very happy with the fact that other researchers that are not physicists, are now becoming interested in the Penna model. We have experience with computer simulations, but are not aware of many important biological aspects that would surely improve a lot the model, if they could be included in the simulations.

Part II
DNA, Money and War

3 Immunology, DNA, and Heartbeats

3.1 Immunological Cellular Automata

Cellular automata (WOLFRAM 1986) consist of binary variables ("spins") ± 1, usually distributed on large lattices. The value of each spin at the following time step is determined by the values of its neighbouring spins and perhaps also of itself at the present time step.

Vertrebate animals have an immune system protecting them against diseases. If a virus or other antigen enters our body, then usually certain white blood cells or antibody molecules produced by them neutralize the dangerous antigen and produce a memory of this disease. When the same type of virus reappears later the immune system remembers it and reacts quicker and stronger. In this way, smallpox vaccination produced an immunological memory, via a controlled and not dangerous disease. Thus the human body was protected in case the dangerous smallpox virus tried to live in it, and since many years the smallpox disease is eradicated. AIDS, on the other hand, is a sickness of the immune system produced long after the infection with the virus (HIV). The review "Immunology for Physicists" by PERELSON and WEISBUCH (1997) explains immunological reality to us physicists; instead we concentrate here, as we do in the rest of this book, on those papers were computer simulation methods similar to statistical physics are employed. If nature behaves differently, it will not get an academic degree from us. In the words of Franco Celada: "in machina" studies should complement the traditional in vivo and in vitro work of immunology.

Immunologists distinguish between humoral and cellular response, and between clonal selection and idiotypic-antiidiotypic networks. We brush over these important distinctions since we look more at the computational aspect. We discuss here only cellular automata models with discrete dynamics (ZORZENON 1998); for differential equations we refer to the review of LIPPERT and BEHN (1997).

The earliest application of cellular automata to immunology seems to be the one of KAUFMAN, URBAIN and THOMAS (1985), and physicists entered the field with WEISBUCH and ATLAN (1988). A wide variety of models has been simulated and is only partially reviewed here, and no consensus is evident from the literature that one of them is a standard model for immunology, like what the Penna model became for biological ageing.

The model of Kaufman et al (1985) with later simplifications uses five types of cells or molecules: Antibodies A, helper cells H, suppressor cells S, white blood cells from the bone marrow B, and the virus V. High concentrations correspond to spin up, low ones to spin down. It is easier to write down the formulas for the change with time, if we regard A, H, S, B and V as logical variables which are true for spin up (high concentration) and false for spin down (low concentration). Then the value at time $t+1$ (on the left-hand side) is determined by the variables at time t (right-hand side of the equation), where the logical operations are shown already in section 2.2:

A = V and B and H
H = H or (V and not S)
S = H or S
B = H and (V or B)
V = V and (not A)

The last equation, for example, tells us that a virus proliferates if and only if there is already some virus present and if no antibodies are present.

This system of only five coupled cellular automata has only $2^5 = 32$ possible states, and it is easy to produce a flow diagram on a small computer (by hand one makes as easily some errors). This trivial simulation shows that there are five fixed points:

naive system: all types absent (low concentration)
vaccinated: helper and suppressor cells present, others absent
immune: helper, suppressor and B cells present , others absent
paralyzed: only suppressor cells present
paralyzed and sick: only suppressor and virus present

If one adds a virus to the system in its second or third fixed point, then the system quickly reverts to the immune state, faster than if we start from the naive fixed point. This basic principle of the immune response is thus simulated by this primitive model. A "fuzzy" generalization with three instead of two states per variable was studied by AHMED (1996).

Another model by WEISBUCH and ATLAN (1988) concerned the special case of autoimmune diseases like multiple sclerosis, where the immune cells attack the nerves of our own body. Again, this model has only five binary variables to describe the whole body. In this case, this and the other models mentioned here thus far refer to only one type of antibody and also neglect any spatial extension of the immune system. These models, both of Weisbuch and

Atlan and of Kaufman, Urbain and Thomas, have also been put on a lattice (DAYAN, HAVLIN, STAUFFER 1988; NEUMANN 1989) as is customary in cellular automata simulations and statistical physics, though not necessarily in real immune systems. Immune sicknesses like AIDS were also simulated in this way (PANDEY and STAUFFER 1989), as well as cancer (CHOWDHURY, SAHIMI and STAUFFER 1991). CHOWDHURY and STAUFFER (1990) let the computer study all possible cellular automata with up to five variables to check which of these have the same fixed points as the original model of Kaufman et al. A problem with some of these approaches is that they postulate the existence of suppressor cells S which is at least not widely accepted in today's immunology.

A probabilistic generalization of deterministic cellular automata was used by PANDEY and STAUFFER (1990) to explain the long time delay between an HIV infection and the outbreak of the full AIDS disease (at present deadly). It uses helper cells H, cytotoxic cells S, virus V, and interleukin molecules I with the dynamics

$$
\begin{aligned}
&V = H \text{ and (not } S) \\
&H = I \text{ and (not } V) \\
&I = H \\
&S = I
\end{aligned}
$$

Thus the interleukin molecules act as messengers from the helper cell to trigger the S cells killing the virus. This model can lead to oscillations of period two between a state with and without S and H cells; the virus is always present. If with some low probability insufficient interleukin is produced by the helper cells in violation of the above rule I = H, then after a strongly fluctuating number of iteration, of the order of the reciprocal failure probability, the model jumps into the fixed point where the whole immmune system is destroyed and only the virus survives. This behaviour has some similarity with reality, when the oscillatory behaviour is identified with the long latency, and the final fixed point with AIDS. Indeed, some clinical experiments helped patients by interleukin administration (BAIER et al 1995). A more unified model incorporating normal immune response, auto-immune diseases, and HIV infection was studied by CHOWDHURY (1993) who reviewed these unifying approaches in CHOWDHURY (1995). A more thorough recent review of AIDS models is given by PANDEY (1996, 1998). (Different probabilistic automata for helper cells were proposed by TOME and DRUGOWICH 1996.)

The various lattice models cited above may not be biologically very relevant since the body fluids are well mixed and a mean field approach with one variable for the whole body (Kaufman et al; Weisbuch and Atlan) may be better. This is no longer the case in the shape space approach to immunology (PERELSON and OSTER 1979) explained now.

Different sicknesses are dealt with by different antibodies and cells corresponding to the particular virus. There may be 10^{11} different types of antibodies which humans can produce; this number is far greater than the number of active genes in our DNA. Presumably with lots of somatic mutations our body constantly tries out new types of lymphocytes (white blood cells) but starts mass production only if these randomly changed cells happen to correspond to some new virus found (clonal selection). The literature estimates for the number of possible cell types seem to increase in time, but so does the power of supercomputers. Thus with the best available computers it is possible to study systems containing nearly the same order of magnitude of cell types as in reality.

The different cell types correspond to different geometric shapes and chemical affinities of characteristic molecules presented on the cell surface. Only a virus complementary to the shape and affinity of the molecules can be neutralized by these antibody molecules. Closest to the bit string ageing model of the preceding chapter (PENNA 1995) are therefore the bit string models of immunology (FARMER, PACKARD and PERELSON 1986) continued by CELADA and SEIDEN (1992) and recently taken up also by CASTIGLIONE, BERNASCHI and SUCCI (1997) and BEZZI et al (1997). If the virus is represented by a string of eight bits which are zero or one, like

1010010

then the complementary antibody is the bit string

0101101

such that their arithmetic sum or logical OR gives 255 = 11111111. Geometrically this means that the position of a hole on the virus surface must correspond to a peak on the antibody surface. Similarly, a key can open a lock only if it is geometrically complementary to the lock, apart from minor scratches. (Actually, only the parts called major histocompatibility complex MHC are relevant.) To have some antibody-antigen reaction, not all bits but most bits (like 7 in the example of 8 bits) must be complementary. The Celada-Seiden automata have virus V, antibody molecules A, helper H and B cells as before, and also plasma B cells and macrophages; the latter ones are unspecific and eat everything, like one of the authors. All these elements diffuse around in a square lattice and interact only locally. The model shows clearly that the immune response to the second infection by the same virus is stronger and faster than that to the first infection. CASTIGLIONE, BERNASCHI and SUCCI (1997) parallelized the computer code and let it run on an IBM SP2 with up to 16 processors working on the problem simultaneously, simulating 512 lattice sites and 4096 different virus types (i.e. bit strings of length 12).

BEZZI et al (1997) further generalized the Celada-Seiden model by introducing three additional aspects: i) The virus may enter into a cell and proliferate there exponentially, protected from antibody molecules, until it kills the cell. ii) Killer cells can kill these infected cells. iii) Cross-regulation allows to switch from an immune response by cells to one by antibody molecules. Their model thus is the most complicated one presented here. They find besides two final states of being sick or being immune also an intermediate oscillatory state, as long as they do not destroy it by a mean field approximation.

The simple yes-no answer for describing the shape space in these bit-string models, and also the simple yes-no method for describing the concentrations in the usual cellular automata merely as low or high, have both been generalized, while still keeping integer arithmetic. The concentrations can be described by integer variables corresponding to the logarithms of the concentrations; thus increasing the variable by one unit increases the concentration by a factor 2.718 (DE BOER, SEGEL and PERELSON 1992; abbreviated as BSP and perhaps a future standard model for idiotypic-antiidiotypic networks). More interestingly, the simple shape space description of holes and protrusions (0 or 1) has been generalized by BSP to allow for L different heights $1, 2, \ldots, L$ in each of d different positions. Height k then is complementary to height $L+1-k$ on the same position. (Instead of purely geometrical properties these numbers may also indicate other affinities.) The shape thus is a point in the d-dimensional hypercubical lattice of L^d sites. And the previous Celada-Seiden model with strings containing 8 bits corresponds to $L = 2$, $d = 8$, i.e. to an eight-dimensional hypercube. And again no complete complementarity is required between the virus and the corresponding antibody molecule or cell; the two points in shape space may have a mismatch of one unit in one direction (nearest neighbour BSP model of STAUFFER and WEISBUCH 1992) or even more (original BSP model). Starting from a random distribution, this model gave circular domains in two dimensions, with a radius proportional to the interaction range. According to PERELSON and OSTER (1979), five to ten dimensions are reasonable for human immunity, and in order to simulate them efficiently the number of possible states for each site was reduced to three (STAUFFER and WEISBUCH 1992), with only nearest-neighbour interactions, giving rise to several variants of the BSP model. Numerically, it is practical to use two separate lattices A and B where sites having the same index in one lattice are exactly complementary to the sites having the same index in the other lattice. In this way more than 10^9 sites were simulated (STAUFFER and SAHIMI 1993), nearly as many as the number of possible shapes in the immune repertoire.

Now the concentration variable in one lattice depends on the sum of the variables in the neighbourhood of the other lattice:

$$S_i^A(t+1) = f(\sum_j S_j^B(t)), \quad S_i^B(t+1) = f(\sum_j S_j^A(t)) \quad . \qquad (3.1)$$

Here the index j runs over i and the nearest neighbours of i. If more than two states of high and low concentrations are possible for each S_i, the function $f(h)$ in the above equation may also depend on the current value of the spin to be updated. Since the existence of suppressor cells is doubted, the experimentally observed log-normal immune response to the virus concentration is approximated by window automata (NEUMANN and WEISBUCH 1992): $f(h) = 1$ if the "field" lies in some intermediate interval, and $f(h) = 0$ (or -1) outside this interval. (For more than two states, S_i is increased or descreased by one unit if h is inside or outside, respectively, of this interval.)

With the window covering roughly the center third of the possible range for h, one may get in five to ten dimensions as a function of the initial conditions a phase transition from few sites with large concentrations, to one where a large fraction of all sites carry a large concentration. For a large fraction of excited sites, the immune system has gone crazy and attacks nearly everything it finds. Only the other possibility of a small fraction of excited sites may correspond to a healthy immune system reacting specifically to the disease it should fight against. Somewhere in between a percolative transition occurs where the excited sites (S_i large) form an infinite connected network coexisting with smaller disconnected clusters. Biologically, only about 15 to 20 percent of the lymphocytes participate in the functional network according to COUTINHO (1989). BERNARDES and ZORZENON (1997) show the clusters appearing in the simulations of their three-dimensional BSP model; these clusters grow and shrink, separate and coalesce in a dynamical equilibrium. (J.G.E. Batista, unpublished, combined the model of Kaufman et al with the window automata ideas and got periodic limit cycles.) PAPA and TSALLIS (1996) combined shape-space immunology with interactions in real space.

Presumably more important is the edge of chaos (BERNARDES and ZORZENON 1997) which separates the region of spreading damage from that of the damage remaining localized or healing. For this purpose one simulates two replicas of the immune model simultaneously. They are identical and use the same random numbers. The only exception is that in one system a single "spin" (or a small region of neighbouring spins) was set up when in the other it was down. In later site-by-site comparisons of the two models, the "damage" is the set of spins differing in this comparison. This damage thus describes the effect which the initial isolated perturbation has on the system. It may go to infinity in an infinite system, it may go to zero, or it may remain finite and positive, corresponding to chaotic, stable, and marginal behaviour, respectively. The whole concept seems to be due to KAUFFMAN (1969) where it referred to the effect of genetic mutations. In immunology it may describe the effect of the attack by virus of one single type on the whole immune system. Also, these damaged regions in the BSP models simulated by BERNARDES and ZORZENON (1997) are not static but approach a stationary dynamical equilibrium.

Continuation of this work (ZORZENON and BERNARDES 1998) could reproduce the effects of ageing on the immune response of mice: At time = 0 four copies 0, 1, 2, and 3 (to avoid an immune reaction from C programmers among readers and authors) of a stationary BSP simulations are made. Copies 1 to 3 are then subject to permanent small challenges of the immune system, achieved by flipping some spins up and representing the normal dangers of life; the damage produced by these flips is evaluated by comparison with the unchallenged copy 0. Then, after 250, 800, and 2400 iterations for copies 1, 2, and 3, respectively, a strong dose of antigen is administered by flipping regions of neighbouring spins to high concentrations. The immune response is measured by the amount of damage in comparison with copy 0. In agreement with experimental studies on mice it turns out that for the young mouse (copy 1) the relative immune response to the strong dose of antigen is much stronger than for the old mice (copies 2 and 3): The BSP model shows signs of age.

In the first simulations it looked as if only above four dimensions such a phase transition is possible (SAHIMI and STAUFFER 1993), also for Ising-type models with only two states per site (DE BOER, VAN DER LAAN, and HOGEWEG 1992). Later work using different parameters showed, however, that also in two dimensions such transitions appear (ZORZENON and BERNARDES 1995; MEYER 1995). In summary therefore, these BSP-type models of the immunological shape space show a wide range of behaviours which can be fitted to the expected strong but limited immune response close to the edge of chaos, but still in the non-chaotic regime.

More precisely, most of these shape-space simulations do not describe the interactions between a virus and the corresponding B-cells, but between complementary B cells (CHOWDHURY, DESHPANDE and STAUFFER 1994 go beyond that). According to Nobel laureate Niels Kai JERNE (1974) there are "anti - idiotypic" shapes in the immune repertoire which are quite similar to the original virus, and thus complementary to the shapes of the "idiotypic" B cells which fight the virus. There may even be anti-anti-idiotypic cell types complementary to the anti-idiotypic ones. These interactions between different white blood cells are supposed to control immune response and immune memory. The original virus (more generally, the antigen) only triggers the actual immune reactions.

3.2 Walks on DNA

Our genetic code (at least most of it) is stored inside our chromosomes, specifically in the polymeric molecules called desoxyribonucleic acid, or DNA. Each DNA molecule is a long linear chain formed by a backbone where four bases, adenine (A), citosine (C), guanine (G) and timine (T) are attached one after the other. The specific order they appear in along the chain (e.g. GAT . TCA. GCT. ...) corresponds to the genetic code. This code is completely known by biologists since a long time: each sequence of three adjacent bases codes for a specific aminoacid used to synthesize the proteins responsible for our life machinery. Those proteins are also linear molecules, each one corresponding to a specific sequence of aminoacids which determines its metabolic functioning. As 20 distinct aminoacids exist, and the possible combinations of three bases among A, C, G or T counts $4^3 = 64$, we have some degeneracies in this code, more than one such combination corresponding to the same aminoacid. The symmetry behind this degeneracy as well as its evolution during the development of life on earth is the target of recent theoretical researches — see HORNOS and HORNOS (1993) and also the editorial comment of MADDOX (1994).

However, our interest here is another: does evolution introduce some kind of correlations along DNA chains? This question has been investigated in recent years, first by LI and KANEKO (1992) and LI (1992). Then, many other people, in particular PENG et al (1993), VOSS (1992), ARNEODO et al (1995 and 1996), studied the same subject. In order to understand their approach, we must first know some features of a classical problem very familiar to physicists: the random walk. Consider a walker along a horizontal X axis, starting from the origin $x_0 = 0$. It walks by discrete unit steps which can be to the right or to the left in a random way. After the first step, the walker can be found either at position $x_1 = +1$ or at $x_1 = -1$, with 50% probability each. Supposing the first case occurred, after the second step the walker can be found at either $x_2 = 0$ or $x_2 = +2$, and so on. The relevant question is: how far is the walker from the origin, after t steps? As the steps are tossed at random, this question must be answered under a statistical point of view, i.e. by taking averages over many realizations of the walker. One way to do this is to perform the experiment by tossing right or left steps with a coin, and measuring the last position x_t after t steps. Repeating this experiment N times, always with t steps, one gets N values of x_t which can be plotted on a histogram: for each distinct value of x on the horizontal X axis, one plots on the vertical axis the fraction of times $P_t(x)$ this value arises from the N realizations. In order to accelerate the process, the experiment can be performed using a computer with (pseudo) random numbers instead of a coin. For large enough N and t, one can verify that the plot is a bell shaped curve centered at the origin. The results of such a simple computer program is shown in figure 3.1. Simple mathematical reasonings show that those curves are Gaussian probability distributions.

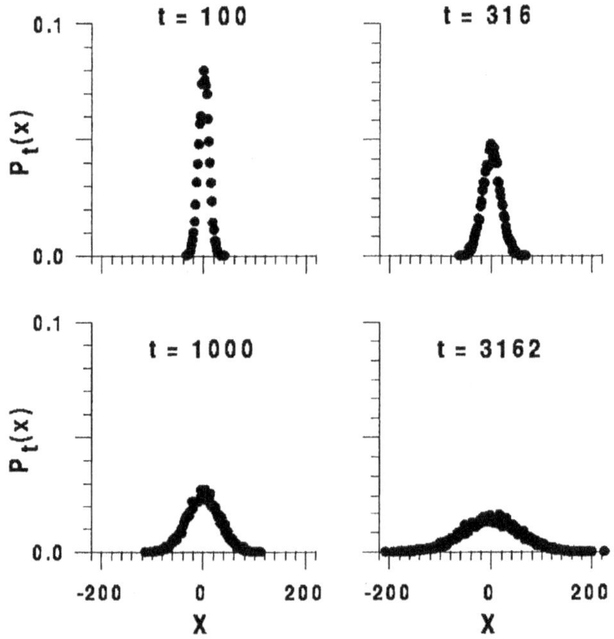

Fig.3.1 - Probability distributions $P_t(x)$ for a one-dimensional random walk, measured for different times t (number of steps). All cases correspond to 10000 samples.

Repeating the same experiment for different values of t, one can also observe the increasing width of the bell shaped distribution curve $P_t(x)$ for increasing values of t (always taking a large enough N). A naive way to estimate this width is to measure it just at the half maximum height appearing at $x = 0$. Mathematically, one can also compute the width Δ_t by the well known variance formulae $\Delta_t^2 = < x^2 > - < x >^2$, where $< q > = \sum P_t(x)q(x)$ represents the average of some x-dependent quantity q over all realizations of the experiment. The smart reader will note that $< x >$ vanishes for large enough N, simplifying the task. However, $< x >$ does not vanish for a finite N, and calculating its value for a specific set of realizations of the experiment is a good way to estimate the statistical errors involved, determining whether N is or is not "large enough". Figure 3.2 shows the dependence of Δ_t on t as a straight line (full circles). Being a $log \times log$ plot, this figure indicates a power law dependence

$$\Delta_t \propto t^\alpha \quad , \tag{3.2}$$

where the exponent $\alpha = 1/2$ can be measured as the slope of this straight line. Simple mathematical arguments show this is indeed the exact value.

The exponent $\alpha = 1/2$ is the signature of the random process we are dealing with. It arises because our walker is completely drunk, and forgets all previous steps before deciding to go further to the right or to the left. The walker has no memory at all. Would our walker have some degree of memory, the already performed steps influencing in some way its decision concerning the next step, the characteristic exponent α would be distinct from 1/2. In this case we call it a **correlated** (not random) walker. Values $\alpha > 1/2$ correspond to persistent walkers that tend to repeat their past history: a drunk guy remembering some previous step, say, to the right, which successfully lead him to an open pub, tends to go again to the right trying to find another open pub. For this case, the width of the bell shaped distribution curve increases with time **faster** than the random walk. On the other hand there are **anti-correlated** walkers which tend to repeat previous steps in the opposite senses, as the alternance between parties in a democracy. In these cases one has $\alpha < 1/2$, and the distribution curve widths increase with time **slower** than the random walker.

PENG et al (1992) decided to measure the exponent α for DNA sequences, in order to answer the above-posed question about the existence or not of correlations introduced by evolution into our genes. Instead of tossing randomly the successive steps of a random walk, they **read** this information sequentially from a large molecule of DNA. Each purine A or G is interpreted as a step to the right, whereas each pyrimidine C or T represents a left step. Here the "time" flows as successive bases A, C, G or T are read one after the other. In order to construct many walks for the same time, necessary to obtain the width of the probability distribution curve, one uses the same DNA sample for

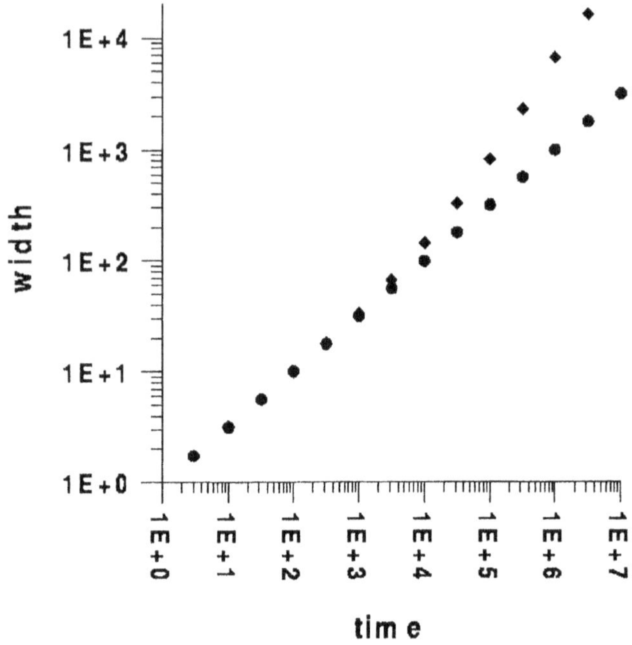

Fig.3.2 - Widths of distributions $P_t(x)$ as a function of time t for the random (full circles) and correlated walks (diamonds).

all, starting each walk from different positions along it. They have performed this task many times for many different DNA sequences corresponding to different species, and have found persistent walker behaviour, i.e. $\alpha > 1/2$ in all cases. Different samples of DNA from the same species give the same value for α, but each species presents its characteristic exponent α. Furthermore, they have found a consistent relation between α and the evolutionary degree: the larger the value of α is, the more evolved is the corresponding species. Algae present values slightly above 1/2, while humans present $\alpha \approx 0.8$, with a lot of values in between for different species. Another interesting point is the possible existence of further correlations due to the three-bases code. CEBRAT and DUDEK (1997) studied this question by skipping the two next bases after the currently read one.

Let's introduce a new ingredient into our random walk, trying to transform it into a **correlated** walk. With a small predefined probability p, say $p = 0.01$, we store in a buffer a small sequence of successive steps, starting from the current one. The length of this small sequence is randomly chosen between 1 and pt. This sequence remains stored until a new one replaces it (this may occur at any time, with probability p). As the growing process continues, with the same probability p for each new step, we insert into our walk the past sequence currently stored on the buffer. The insertion starts at the current position, and then the walk restarts its random growing process after the inserted piece. In this way, some repeated sequences appear along the walk. After t steps (including the inserted ones) the walk is finally done, and we repeat the same process N times, in order to plot the probability distribution histogram. It is again a bell shaped curve, but its width is larger when compared with the random walk case, for the same time t. Moreover, the larger the value of t, the larger the difference between the widths of the correlated and the random walk. Figure 3.2 shows both cases, where a net deviation from the random walk exponent $\alpha = 1/2$ (full circles) can be observed in the correlated walk data (diamonds), for large enough values of t (this simple rule is only illustrative, and the range of correlations introduced is indeed finite). The lesson we can learn from this simple example is: inserting copies of some pieces taken from a random sequence into random positions of itself, the whole sequence may become no longer random, presenting long-range correlations due to the repetitions. The degree of these correlations can be estimated by measuring the exponent $1/2 < \alpha < 1$ through the slope of $log \times log$ plots of the kind presented in figure 3.2.

PENG et al (1992) proposed a theory to explain the correlations found in DNA sequences. Living beings reproduce themselves, but the offspring are not exact copies of the corresponding parents, due to mutations (and crossings whenever recombination is present). These mutations occur mostly during the reproduction process, when the DNA molecules are duplicated by a set of specialized enzymes, inside the parent's cells. One already existing molecule of DNA serves as template in order to synthesize a copy of itself. One extreme of such a specialized enzyme "reads" the bases from the already existent DNA molecule, one after the other, while the other extreme synthesizes the copy at the same time. Sometimes an error occurs, the corresponding basis of the copy being different from the template. These errors are called **point mutations**. DNA molecules are very long linear objects appearing completely folded inside the cells. During the synthesis of a copy, the specialized enzymes unfold the piece currently being copied in order to follow basis after basis the correct sequence. However, sometimes this unfolding process fails, and entire pieces of adjacent bases along the original DNA molecule are not copied at all. These mutations are called **deletions**. Sometimes the special enzymes return back to an already "read" point along the DNA molecule, and entire pieces of adjacent bases appear twice on the copy, at positions which can be far apart from each other. These are called **insertions**, and are just the new ingredient we have introduced into our random walk example, transforming it into a correlated walk for which the exponent α becomes larger than 1/2. Thus, one can understand why the values of α measured in real samples of DNA are always larger than 1/2: it is a consequence of evolution, after many, many generations. Also, the more evolved is the species under study (the larger its past evolution time, including ancestor species), the more likely are insertions. Thus, the larger values of α are characteristic of the more evolved species (like German Herr Professors, compared with Brazilian teachers), as found by Peng et al.

Along DNA molecules there are entire pieces of adjacent bases which are not used to code proteins. These pieces are called **introns**, whereas the other DNA regions known to code some protein synthesis are called **exons**. There is a huge controversy among biologists about the function of introns: no consensus is quite evident for the reason of their existence. For humans, around 2/3 of the whole set of DNA molecules correspond to introns. For algae, introns are almost absent, and the whole DNA repertoire codes for protein synthesis. The more evolved is the species, the larger is the fraction of (useless ?) introns present in its genetic code. Skipping the introns, i.e. reading only the coding pieces of DNA samples, Peng et al found $\alpha = 1/2$ in all cases, indicating that long range correlations appear only because introns are present. The actual coding regions of our genetic code do not present long range correlations. Peng et al proposed that introns are fossil, no longer active parts of our genetic code. In some point of the evolution time, some piece of the DNA was erroneously inserted in another position: two copies of the same genetic piece code for the same protein synthesis. After many generations, one of these copies may suffer a deleterious point mutation, becoming useless: hereafter this piece remains as an intron. However, it is not much different from the other copy still coding for protein synthesis, and this similarity would be the responsible for the observed long range correlations.

This beautiful theory must be tested by experimentation, according to the untouchable dogma of science development. However, it is very hard to perform experiments on biological evolution. Normaly one is able only to observe the consequences of many past evolutionary processes in the current population. In order to perform experiments, one would be able to repeat the evolutionary history of the species under study. The time scale involved in these potential experiments are normaly some orders of magnitude larger than the life span of the present authors, in spite of their longevity. We are trying to avoid ageing through our computer simulations presented in chapter 2, but without success until now. Then, instead of experiments, we would prefer computer simulations also for the problem of DNA long range correlations.

As one of the authors, I am now (late night) typing these lines on my computer. The two lazy associate professors, coauthors of this book, ask me to go home, we are all tired. In a few minutes I will type a "save" command, and these lines will be stored on my computer disk as they are. Tomorrow morning, I will read again the whole thing: certainly my above mentioned enemies will find some corrections to do, new material to include, etc. After all this stuff, I will type another "save" command, and the new version will again be stored on my computer disk. Would the new version be stored in exactly the same disk position as the old one? Not necessarily, because the new version may need more (or less) disk space than the old one. The operational system itself chooses the best position to store the new version, according to its own pre-programmed criteria. Probably some portion of the old version will be lost, the disk space formerly occupied by it being now used to store another portion of the new version. On the other hand, some other portion of the old version will not be touched, the operational system simply marks the corresponding disk space as free for future use, without erasing anything there. Its contents remain on the disk as fossils, until the operational system decides to replace it by something else. Let's call these fossil informations by the name introns: they are pieces of old file versions which still remain stored on the disk (in an idealized infinite disk, they would remain forever there). Normally, the user has no access to these informations, but there are some softwares which allow one to read them (for instance, the softwares designed to "recover" files which were erroneously deleted). On the other hand, parts of the disk occupied by currently active files will be called exons.

ZEBENDE, PENNA and OLIVEIRA (1998) performed an experiment using diskettes in order to test the theory proposed by Peng et al. First, the diskette is completely filled with random bits 0 or 1, with 50% probability each. Reading these bits one after the other, one can construct a random walk and measure the exponent α. The result, of course, is $\alpha = 1/2$ within the statistical error bars. Now, we store some files on this diskette, whose bits 1 or 0 are again taken at random with 50% probability. The exponent $\alpha = 1/2$ is again measured. Then, we perform a random edition on each file. This random edition consists of two steps. First, we randomly modify a fraction, say 10%, of the bits on the file (point mutations). Second, we include some new small random sequence of bits on the file (insertion) or, alternatively, we delete some small sequence from the file (deletion). After this random edition, introns appear. Measuring now the exponent α, one gets a value slightly larger than 1/2. In that measure, the whole diskette must be read, one bit after the other, disregarding which parts of the diskette correspond to the current file versions, and which do not. Now, we perform again another set of random editions on the current files, another one, and so on. The measured values of the α exponent increase while we perform more and more editions (evolution time), eventually saturating in some value between 1/2 and 1 which depends on the diskette and file sizes, the number of files, etc. Many experiments were performed, with different numbers or sizes of the files, including or not backup copies, varying the size of the diskette, etc. Including backup copies, for instance, larger values of α were found. By reading not the whole diskette, but only the bits belonging to currently stored files, i.e. reading only exons and skipping introns, one finds $\alpha = 1/2$ in all experimented cases.

These results obtained from a simple simulation of evolution using diskettes completely agree with the observed DNA behaviour, as well as with the theory proposed by Peng et al. It is a nice example of the power of computer simulations in general. One of its advantages (not, by far, the only one) is the possibility of performing "experiments" where real experiments are impossible or very hard to implement.

3.3 Heartbeats and Leaky Faucets

A very simple and interesting experiment was first performed by SHAW (1984). It consists in measuring the times between successive drops falling from a leaky faucet. One uses a laser beam detected by a photocell: when the drop passes through the beam, deflecting it away from the photocell, an electronic signal is sent to a computer which stores the time spent since the last drop. After many successive drops, one has a time series to study. Are these numbers (time intervals between successive drops) random? If not, has the dripping faucet some kind of long range memory, the current time interval depending upon all past drops? These questions are very similar to that presented in section 3.2 concerning the DNA sequences. To answer them, PENNA et al (1995) constructed walks from the experimental sequences of numbers, and measured the exponent α. This procedure was already well discussed in section 3.2. Recently, BONABEAU et al (1998) showed that living ants may behave like water molecules on the leaky faucet.

Before discussing the walks, however, we can observe the behaviour of the time series through the so-called return maps. Let's call $T_1, T_2, T_3 \ldots$ the series of time intervals between successive drops, taken from the experiment. Plotting T_{n+1} versus T_n we get the (first) return map. The simplest conceivable case is a perfectly periodic faucet, for which all time intervals are equal. In this case, the return map consists in only a point located at the main diagonal of the first quadrant. As this ideal case never occurs due to the existence of noise, one observes actually a small shadow of points located inside a small region centered in some point of the quoted diagonal. We will call this case quasi-periodic. One can also plot the probability $P(T)$ of measuring the time interval T, by dividing the time axis in small intervals, and counting how many times each interval is visited during the experiment. This probability distribution is again a bell shaped curve similar to the ones discussed in section 3.2 for the DNA sequences. The width of this distribution curve corresponds to the noise. In the ideal case of a perfectly periodic faucet, the width would vanish.

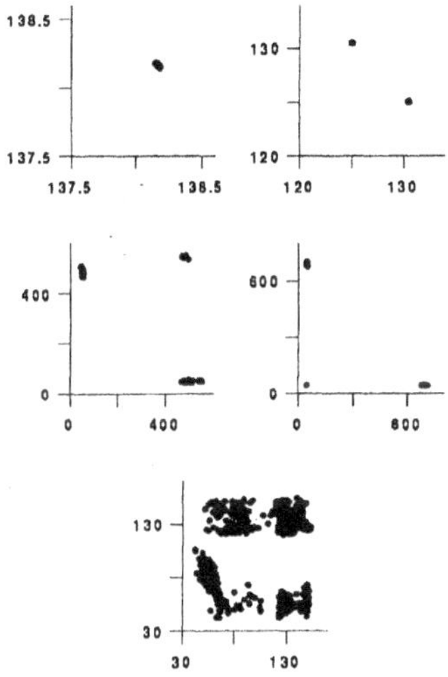

Fig.3.3 - Experimental first return maps $T_{n+1} \times T_n$ for successive drops falling from a leaky faucet. The experiments were kindly performed for this book by J.C. Sartorelli and R.D. Pinto in University of São Paulo. Different plots correspond to different water fluxes.

By controlling the constant flux of water into the faucet, one can observe the quasi-periodic case, as well as many other patterns, some simple and others very complex. Another very common pattern actually observed corresponds to the appearence of small satellite drops. Each large drop is followed by one small drop (large, small, large, small, large, ...). The time intervals between successive drops will also follow such a pattern (large, small, large, small, large, ...), once the water flux is constant and the time interval must be

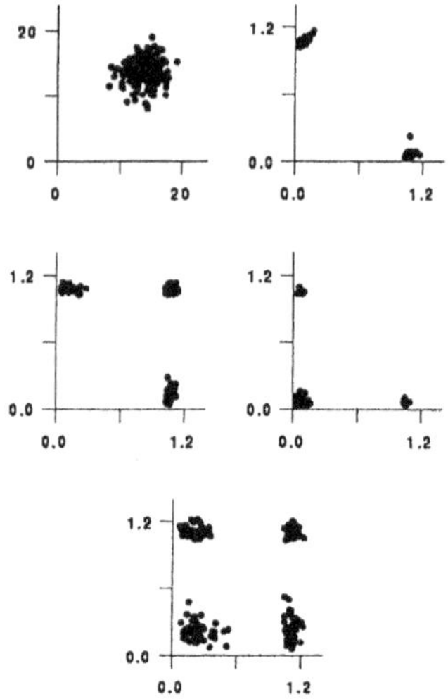

Fig.3.4 - Computer simulated first return maps $T_{n+1} \times T_n$ for successive drops falling from a leaky faucet. The simulations were kindly performed for this book by A.R. de Lima in University Fluminense, Niterói.

proportional to the drop mass. The return map for this pattern consists of two small shadows symmetric with respect to the diagonal of the first quadrant. A more complex pattern corresponds to the case where the satellite small drop sometimes appears after a larger one, but sometimes it is missing (large, small, large, large, small, large, large, large, small, large, etc). In this case, the return map presents three shadows, two of them symmetric with respect to the diagonal, and the third one at the diagonal. Another pattern corresponds to the possibility of more than one satellite following a large drop (large, small, small, large, large, large, small, large, ...). The return map is a square of four shadows, two at the diagonal and the other two symmetric with respect to it. Other more complex patterns are also observed, depending on the constant water flux controlled by the experimenter. Figure 3.3 shows some examples of real drops taken from experiments performed at University of São Paulo — see SARTORELLI, GONÇALVES and PINTO (1994), PINTO et al (1995), and also MARTIEN et all (1985), YÉPES et all (1989), WU and SCHELLY (1989) and DREYER and HICKEY (1991). Figure 3.4 shows similar results taken from computer simulations performed at Fluminense University in Niterói, Rio de Janeiro — see DE OLIVEIRA and PENNA (1993), (1994), PENNA et al

(1995) and DE LIMA, PENNA and DE OLIVEIRA (1997). The simulational model will be treated later in this section.

The experimental group from University of São Paulo used also another way to observe the drops, by recording the successive drop falls on video tapes. This allows also to study the morphology of a single drop, besides the dynamics of a sequence of many drops. Various quantities as, for example, the drop volume or its center of mass height have been measured as a function of time. Figure 3.5 shows these results for both experiments (left) and simulations (right). Other quantities also present the same behaviour in both experiments and simulations. Once we have a simulational model which reproduces very well the behaviour of real falling drops, both within the time scale of many successive drops as well as the much smaller time scale of a single drop, we can try to use this model to go further, studying the system in more details.

Let's introduce the simulational model. Consider a square lattice where a single connected cluster of Ising spins pointing up represents the drop, surrounded by Ising spins pointing down, representing the air. Each spin interacts with its eight nearest and next nearest neighbours, along both the lattice axis and the diagonals. The interaction is ferromagnetic, i.e. each pair of neighbouring spins pointing in opposite senses contributes with a positive value (taken as the energy unit) to the total energy of the system. Parallel spins do not contribute to the total energy. Two different physical ingredients are present in this spin pair interaction. First, inner spins pointing up tend to stay together, keeping the drop connected, because this minimizes the total energy (parallel neighbours do not contribute). This mimicks the van der Waals attraction between water molecules, in a real drop. Second, all the energy is distributed along the drop surface, where one has spin pairs pointing in opposite senses: to minimize the energy corresponds thus to minimize the drop surface. This mimicks the surface tension on real drops. The third physical ingredient one needs to include is the earth gravitational attraction pulling the drop downwards. Besides the pair interaction with its eight neighbours, each spin also interacts with an external magnetic field decreasing (increasing) the local energy for spins pointing up (down). This field is uniform along each horizontal row of the square lattice, but increases from above to below (according to a uniform gradient g) along the vertical direction. This field produces an energy gradient just like the earth gravitational potential energy, proportional to the height above the ground.

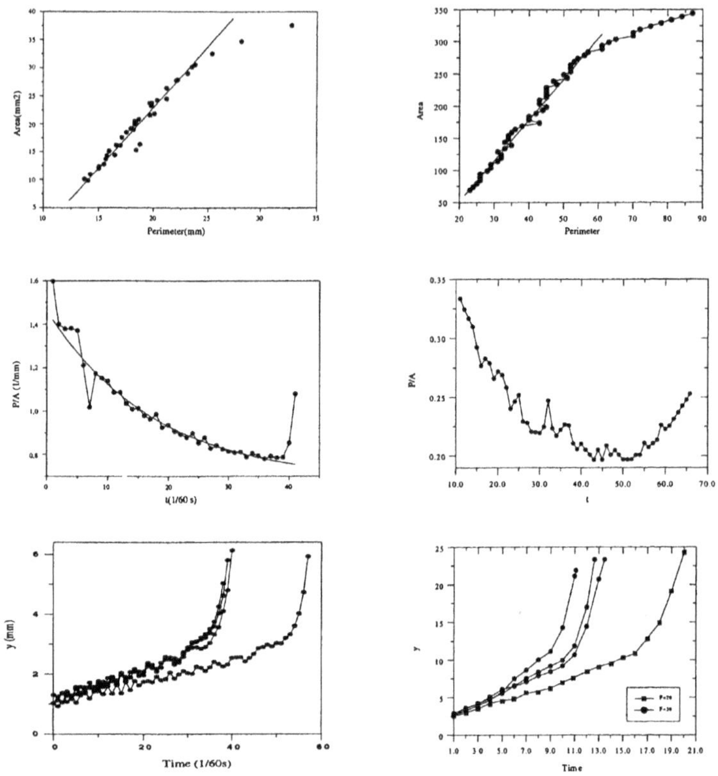

Fig.3.5 - Center of mass height and surface/volume ratio as a function of time, and surface as a function of volume for a single experimental (left) or simulated (right) falling drop.

This simulational model was first introduced by MANNA, HERRMANN and LANDAU (1992) in order to determine the geometry of a drop sticking on a wall, a static optimization problem, where the drop mass is conserved. The drop relaxes through a Kawasaki dynamics where one spin up and another down are flipped simultaneously. These two spins are randomly chosen along the current drop surface, and the flipping of both is accepted whenever the total energy decreases (thermal fluctuations can be included through a very small temperature, using Metropolis acceptance rule, but they are not important at all). This relaxation process is repeated many times, by tossing two other spins along the surface, and so on. DE OLIVEIRA and PENNA (1993) and (1994) extended this model to the dynamical behaviour of the leaky faucet, now a real dynamic problem. The faucet is a set of W adjacent spins located at the upper row of the square lattice, always pointing up. The drop mass is no longer constant, but it increases linearly with the time, due to the constant flux of water injected into the faucet. This mass injection is performed — see DE LIMA, PENNA and OLIVEIRA (1997) — by flipping F randomly chosen spins pointing down, all belonging to the current outer drop surface. This injection of mass is repeated after a set of V successive mass conserving relaxations of spin pairs (described above). Thus, the drop evolves by relaxing its surface V times, suffering one injection of F new unit masses, relaxing V times again, followed by another mass injection, and so on. The parameters of this model are the faucet width W, the "gravity" constant g, the mass flux F and the number V which represents the fluid viscosity.

At every step, one verifies if the current drop is still a single connected piece of spins pointing up, or, alternatively, if there is some piece disconnected from the faucet during the last movement. Would the latter case occur, we interpret this as the drop fall: the piece disconneted from the faucet is erased, and the time elapsed since the last fall is stored into the computer memory for further treatment. Then, a new drop starts to be formed from the remaining piece still connected to the faucet. This is the crucial step of the simulation, giving raise to a long term memory: the new drop will be formed starting from the part of the last one still connected to the faucet. Thus, one drop "remembers" the previous one, which "remembers" its own predecesor, and so on. This memory mechanism is responsible for the long range correlations which appear, as we will see now.

PENNA et al (1995) constructed walks from the series of numbers (time intervals between successive drops) obtained from both the experiments and simulations. First, considering a given series of numbers $T_1, T_2, T_3 \ldots$, all positive, one takes the average value T, and construct another series $s_1, s_2, s_3 \ldots$, where $s_n = T_n - T$ may be positive or negative, the average of all s_n vanishing. The positive and negative numbers of this new series are the steps of the walk. The walker starts at the origin $x_0 = 0$ of the X axis. After one time step it can be found at $x_1 = s_1$, after the second time step at $x_2 = s_1 + s_2$, and so on. After t steps, one has $x_t = s_1 + s_2 + \ldots s_t$. In order to construct a distribution probability curve, one needs many walks, as already described for the random walk. This is obtained from the same series, as in the DNA case, by starting the second walk from s_2 instead of s_1, i.e. $x_0 = 0$, $x_1 = s_2$, $x_2 = s_2 + s_3$ and $x_t = s_2 + s_3 + \ldots s_{t+1}$. The third walk starts from s_3, and so on. For this, of course, one needs a long series, allowing to average over a large number N of walks, i.e. $N >> t$. The smart reader certainly realizes one needs to construct only the first walk. Instead of constructing the second, it would be much faster to add $s_{t+1} - s_1$ to the value x_t already obtained for the first one. For the third walk one further adds $s_{t+2} - s_2$, and so on.

The exponent $\alpha = 0.00$ (with the precision indicated by these digits) was measured in all cases, both from experimental and simulational data. A large amount of series were tested, under different conditions, corresponding to different return maps, all of them giving the same value for α. As one has $\alpha < 1/2$, the leaky faucet is an anti-correlated dynamical system — see section 3.2 — the width of the probability distribution curves increasing with time slower than the random walk. The particular value $\alpha = 0$ means that either these widths do not increase at all for large enough times (this is indeed the observed behaviour), or that they (could) increase very slowly (say, proportional to $\log t$). This corresponds to the most extreme case of anti-correlations. It is not a surprising behaviour for the leaky faucet, once the flux was kept fixed during the experiments or simulations. However, further simulations were performed in Niterói by Adriano Roberto de Lima, allowing the flux F to fluctuate around some average value, and the exponent $\alpha = 0.00$ persists.

Would the measured value for α be equal to $1/2$, then the time intervals between successive drops would be random, without correlation along the series. Does the true value $\alpha = 0 \neq 1/2$ means that they are not random at all? In order to analyse this point, PENNA et al (1995) measured again the value of α for the same series obtained from experiments and simulations, sorting a random order for the numbers. Instead of taking the values $s_1, s_2, s_3 \dots$ following the chronological order they appeared in the experiment or simulation, first they sort the same numbers according to another random order, say $s_{2845}, s_{372}, s_{817} \dots$, and then measured the value of α again. The result is $\alpha = 1/2$ for all cases, within the error bars. One concludes that the long range correlations responsible for the measured value $\alpha = 0$ are due to the true chronological order by which the drops fall one after the other, and not due to the numbers in the series themselves.

Let's look now at the probability distribution curve $P(s)$ for these numbers. It is obviously independent of the chronological order they appeared in during the experiment or simulation. Consider only the simplest case of quasi-periodic faucets. Are these numbers really random? Are correlations completely absent? If so, this curve would neccessarily be a Gaussian distribution, like the random walk. PENNA et al (1995) try to fit Gaussian distributions to the data obtained in these cases, again taken from both experiments and simulations. The fits were not satisfactory at all: the "tails" of the bell shaped curves are much more extended than the Gaussians supposed to fit them. Gaussian curves present fast exponentially decaying tails, whereas the observed curves seem to present slower decaying power law tails. One can conclude the series of time intervals for successive drops present another extra, independent kind of correlation, besides the chronological ordering correlation already detected through the exponent $\alpha = 0 \neq 1/2$.

Lévy probability distribution curves are defined through an integral mathematical formulae. This formulae is so ugly that we will not reproduce it here, inside a so beautiful book (one of my co-authors asks me to mention also his own beauty). Two important points about Lévy distributions deserve our comments. First, they apply to abnormal walks where some enormous steps (the so-called Lévy flights) have a finite probability to occur. As a consequence, such a distribution curve presents tails decaying with power laws. Second, inside the ugly integral formulae there is an exponent ψ which can assume values between 1 and 2, depending on the real case under study. A Gaussian distribution is the limiting case $\psi = 2$, whereas a Lorentzian distribution corresponds to the other limit $\psi = 1$. In between these two extremes, the value of ψ determines the class of problems we are dealing with, in particular how slow is the power law distribution decaying on the tails. The many quasi-periodic leaky faucet data are very well fitted by Lévy distributions corresponding to $\psi = 1.75 \pm 0.1$. This value, independent of the other $\alpha = 0.00$ exponent, shows there is some sort of correlations in the series of time intervals between successive drops falling from a leaky faucet, even ignoring the chronological order in which they appear.

PENG et al (1993) and STANLEY et al (1994) performed another interesting experiment, by considering the time intervals between heart beats of a human being. They measured exactly the same exponents α and ψ discussed above concerning leaky faucets. They also found $\alpha = 0.0$ (with a poorer precision than for the case of the leaky faucet), and $\psi = 1.7$. The coincidence of both these exponents strongly suggests that leaky faucet and heart beat dynamics are governed by the same underlying physical mechanism. At this point, we are entering into the dangerous but fascinating exercise of speculation. Let's do that. Some common characteristics of both heart beats and leaky faucet dynamics can be easily detected. In both, two very different time scales compete. First, a slow, continuous **charge** period occurs, when the drop grows before falling, or when the heart continuously receives blood from the lungs. Second, a much faster process of **discharge**, when the drop breaks and falls, or when the heart contracts itself injecting blood into the whole body. This **slow charge** versus **fast discharge** alternance is one of the main characteristics supposed to be responsible for the so-called self-organized criticality introduced by BAK, TANG and WIESENFELD (1987) — see also BAK and CHEN (1991); BAK 1997 — and found in many different natural dynamical systems since then. These systems control themselves automatically, staying always in a critical dynamical state where long range correlations can be observed.

The most famous example is the Gutenberg and Richter law for earthquakes: if one classifies earthquaques according to their intensities, and plots the number of them versus the intensity, the result is a decaying power law. This means that the underground movements of tectonic plates is a critical phenomenon. Indeed, these plates move very slowly one against the other, during the charging process. Suddenly, the accumulated stress is discharged in few seconds, during the earthquaque. The decaying power law distribution is a terrific reality in this case. It means that one has not a **typical** earthquake intensity limiting the possible range of their destructiveness. For an exponentially decaying distribution, for instance the emission rate of a radioactive sample decreasing exponentially with the time, one has a typical value beyond which the probability of occurrence is extermely small. In the radioactivity example, this typical value is the average life time, after which the sample is no longer dangerous. Mathematically, however, the same average could be infinite for a power law decaying distribution. Thus, in the earthquake case, one cannot neglect the possibility of a huge one, occurring tomorow, which will destroy the west cost of North America or the surroundings of Tokyo. Similar behaviour was also found by analysing weather data — see KOSCIELNY-BUNDE et al (1996/8).

Another important example, although with much less quantitative evidence, is the Darwinian evolution of species. If one classifies the different species according to their degrees of evolution, and plots the number of species versus the corresponding degree, the result would be also a decaying power law. We used "would be" instead of "is" because of two technical problems. First, one needs a quantitative definition of "degree of evolution". In the absence of a better definition, one can use the total genome length, measured by the total number of bases A, C, G and T present in the DNA molecules. Some very few species, however, present genomes much longer than others which can be classified as equivalent to them by a simple visual inspection. Following a strong and old tradition widespread among physicists, one can disregard these pathological cases as "points out of the curve", for which there is an "evident, unimportant error", closing the question. Second, a more difficult problem concerns the number of known, already classified species, which is surely much less than the true number of species living now on earth. Worse than that, surely our ignorance about unknown species is not equally balanced among the more or less evolved ones: the small subset of species already classified by humans is not a good, representative part of the whole set. In spite of these problems, the power law distribution of species according to the evolutionary degree is a very plausible hypothesis, seriously taken by some serious biologists, paleontologists, etc — see, for instance KAUFFMAN (1991), (1993) and (1995). They have many other arguments in favour of this hypothesis. A fascinating presentation of fundamentals of evolutionary theory was given by DAWKINGS (1987), with also a shortened version (1995). Following the same ideas, one can understand the so-called punctuated equilibrium: life on earth evolves by bursts of new species appearing simultaneously, after long periods of stability; also, huge extinction events occur simultaneously at some unpredictable times — see for instance GOULD (1989), BAK (1997).

Could the heart beat dynamics be classified as critical? It is also a plausible hypothesis, since this gives advantages in what concerns the adaptation capabilities of the individum, faced to new, yet unknown challenges imposed by the environment — see PENG et al (1993) and STANLEY at al (1994). Supposing this is indeed the case, how can one explain why a leaky faucet, an obviouly non-evolving dynamical system, behaves alike? Anyway, there is a precise criterion to decide whether a dynamical system is critical or not: its Lyapunov exponent λ. Suppose you have a dynamical system where a single variable x evolves according to the dynamical rule $x_{t+1} = f(x_t)$, determining its future value x_{t+1} from the current one x_t (f is some known mathematical function). Starting from some initial seed x_0 at $t = 0$, one can follows its evolution by applying the function one time, getting x_1. Now, considering $t = 1$ and applying the same function again, one gets x_2, and so on, determining the whole series $x_0, x_1, x_2 \ldots$. The question to be answered is what will occur if one repeats the whole process again, starting now from a slightly different seed x'_0 instead of x_0. Following the same procedure, one gets a new series $x'_0, x'_1, x'_2 \ldots$. What occurs with the initially small difference $\Delta x_0 = x_0 - x'_0$ as time goes by? Do the successive differences ($\Delta x_1 = x_1 - x'_1$, $\Delta x_2 = x_2 - x'_2$, and so on) increase forever? Or do they vanish after some time?

Let's suppose the difference Δx_t depends exponentially upon the time t, i.e. $\Delta x_t \propto \exp(\lambda t)$ where λ is the quoted Lyapunov exponent (this is, indeed, its definition). One can measure the value of λ for the specific dynamical evolution at hands. This is an easy task, provided the function f is known. If one obtains $\lambda > 0$, the successive difference Δ_t increases exponentially with time, i.e. it explodes. This case corresponds to what scientists call **deterministic chaos**. It corresponds also to an unpredictable dynamical system (although deterministic), because any small uncertainty Δx_0 the experimenter has on the initial condition will fast be transformed into a huge uncertainty as time goes by. The weather on earth behaves like this: in spite of the huge amount of money invested in weather forecasting, nobody is able to predict rain or sunshine for one week from now. The other possibility is to measure $\lambda < 0$. In this case, the system is very simple: two different initial conditions x_0 and x'_0 initially follow different trajectories, but quickly collapse into only one. This case corresponds to a predictable dynamical system: slightly different initial conditions end at the same final situation. Positive or negative Lyapunov exponents, thus, divide dynamical systems in two classes: chaotic ($\lambda > 0$), predictable ($\lambda < 0$). The marginal case ($\lambda = 0$) is the one we are interested in, and has already been commented in section 3.1 concerning the immune system.

In the preceding paragraph, we have supposed an exponential relation between Δx_t and t, leading to the quoted explosive behaviour (for $\lambda > 0$). Indeed, this is only the leading term, other less explosive ones could also be present, i.e. $\Delta x_t \propto \exp(\lambda t) + \ldots$, where the symbol $\ldots$ represents these further, less diverging terms. As the exponential dominates the explosive behaviour, one can simply neglect the other terms. For $\lambda < 0$, the exponential represents the slowest decaying term, other faster ones are included into the $\ldots$, and neglected. However, these further terms cannot be neglected in the special case $\lambda = 0$. This is just the critical situation we are interested in. For $\lambda = 0$ the exponential term is just a constant, being dominated for large times by any other increasing function, no matter how slow it is. Thus, if the Lyapunov exponent of some dynamical system is $\lambda = 0$, this fact does not inform us too much about its actual dynamical behaviour. For instance, one can have a power law dependence such that $\Delta x_t \propto t^\beta$, with a positive β: in this case the distance Δx_t diverges for large times, like a chaotic system, very different from a predictable one. On the other hand, for a negative β two distinct trajectories eventually collapse into only one, like a predictable system, very different from a chaotic one. Nevertheless, the same Lyapunov exponent $\lambda = 0$ holds for power laws with both positive or negative values of β, and cannot distinguish them. For this $\lambda = 0$ case a finer tuning is necessary. This was recently done by TSALLIS, PLASTINO and ZHENG (1997), through a generalization of exponential Boltzmann factors; see LYRA 1998 for a review.

Exponential explosions ($\lambda > 0$) and power law divergences ($\lambda = 0$ with $\beta > 0$) are qualitatively similar: for both, two trajectories initially very near from each other become far apart after some time. Quantitatively, however, they correspond to very distinct behaviours. The exponential explosion is very fast, while the power law divergence is slow. The same comparison applies to exponential ($\lambda < 0$) versus power law ($\lambda = 0$ with $\beta < 0$) decays: in both cases the system converges to a final, predictable state. However, there is a typical waiting time for convergence in the case of exponential decay, namely $\tau = -1/\lambda$, whereas one needs to wait an infinite time in the case of a power law decay. This quantitative distinction between exponentials and power laws is a crucial point in the study of critical phenomena. There is an instructive example emphasizing this crucial point. Consider the so-called logistic map

$$x_{t+1} = a\, x_t(1 - x_t) \ , \tag{3.3}$$

where a is a constant control parameter fixed at the beginning. Let's consider $0 < a < 1$. One starts from some seed, for instance $x_0 = 0.1$, and numerically iterates this equation repeatedly. After few iterations one gets $x_t = 0$ and sticks there: this is the final, stable value of x_t. We strongly recommend the reader to do this numerical experiment, by programming equation 3.3 on a computer or pocket calculator. Any other seed x'_0 inside the unit interval (i.e. $0 < x'_0 < 1$) will have the same destiny, sticking at the fixed point $x_t = 0$ after some few iterations.

The mathematical minded reader will recognize that equation 3.3 can be translated into the differential equation

$$\frac{dx}{dt} = -\,(1 - a)x - ax^2 \ . \tag{3.4}$$

As time goes by, as already discussed in the preceding paragraph, x vanishes, thus, the second term involving x^2 in equation 3.4 can be neglected, and the solution of the differential equation is the exponential decay

$$x \propto \exp[-(1 - a)t] \tag{3.5}$$

(remember we are restricting ourselves to the case $0 < a < 1$). This explains the fast convergence to $x = 0$ the reader has found in her/his numerical experiments. The characteristic waiting time for convergence is $\tau = 1/(1 - a)$, i.e. in order to get convergence one needs to iterate equation 3.3 until some t equal to, say, ten times τ (of course, the precise value depends on the numerical accuracy of the machine). Thus, the nearer to 1 is the control parameter a, the longer is the waiting time τ, as can be easily verified numerically.

Let's then extend a little bit our restriction, allowing to choose $a = 1$. The reader will no longer get the correct fixed point $x = 0$, no matter how many interation she/he has programmed her/his machine to perform. The numerical result is simply **wrong**: for a 16-digit precision machine, for instance, the last 8 digits are not zero after convergence. Instead, some seed-dependent wrong digits appear there, in the format $x = 0.00000000ABCDEFGH$ and they have nothing to do with the dynamical system itself. For this $a = 1$ case, the first term in the right-handed side of equation 3.4 vanishes, and one can no longer neglect the second term as we have done before. As a consequence, the solution of equation 3.4 is no longer the exponential decay 3.5. The new solution is

$$x \propto t^{-1} \ , \tag{3.6}$$

now a power law decay. With $a = 1$, we are facing a critical dynamical system, and the concept of waiting, or typical time is lost: one can wait until eternity and the machine will not give the correct result. It is not a matter of the limited machine precision. For an ideal, infinite precision machine each new iteration will approach the current $x = 0.000\ldots000\text{ABCD}\ldots$ a little bit closer to the final value $x = 0$. Here, ABCD... represent the digits not yet converged. However, the effect of the new iteration does not occur at the leading digit A, but at another one located far to the right (namely, two times more distant from the leftmost 0 digit than A itself). Physicists call this slow convergence (more precisely, no convergence at all) **critical slowing down**.

I am sure the reader has performed the proposed numerical experiments with extreme dedication (the reader who fails to do that is no longer reading this chapter, and would prefer to read other parts of the book, written by the other, lazy authors). Thus, the dedicated reader can also try other more exciting values of a, for instance, $a = 3$ and its neighbourhood. In order to observe the convergence, for $a > 3$, one needs to compare the current x_t not with its predecessor x_{t-1}, but with the previous one x_{t-2}. Another interesting point is $a = 3.449489743$. Another yet is $a = 3.5440903596$, and so on. All these values correspond to critical dynamics, the Lyapunov exponent being $\lambda = 0$ in all cases. In between these values, one has predictable dynamics with $\lambda < 0$. Beyond another critical value $a \approx 3.57$, one can find also chaotic behaviour.

After this long digression about exponential versus power law behaviour, let's return back to our heart beats and leaky faucets. Are these two dynamical systems critical? We do not know. It is not an easy task to determine the Lyapunov exponent in this case. We do not know which is the mathematical function (such as that in equation 3.3) determining the time interval T_{n+1} from the previous value T_n. We know only the series $T_0, T_1, T_2 \ldots$ obtained from the experiment or simulation. How to restart the system again from some new seed T_0' different from but very near the original one T_0? There are a lot of numerical methods, some very sophisticated, to perform this task. Unfortunately, they seem to be not very good for critical cases. Applying them anyway, Reynaldo Daniel Pinto from University of São Paulo and Adriano Roberto de Lima from Niterói found $\lambda \sim 0$, but with a very unsatisfactory precision for the experimental series. For the simulated ones, the situation is a little bit better, but even so we are not yet sure that λ is indeed null. Until now, we can only affirm that $\lambda = 0$ is not ruled out by the experimental or simulational data. Work along these lines is still in progress.

3.4 Evolution of Bird Songs

There is no consensus about the definition of complex systems. Some fundamental characteristics, however, are common to most examples normally classified as complex in the specialized literature. First, they are systems composed by many units linked to each other through a web of conflicting mutual influences. Second, such a system evolves dynamically, the destiny of each unit decided (anew every time) according to those influences currently acting on it. What normally one is interested in about these systems is the fact that they usually present power laws, and critical behaviour. In other words, such a system usually lies on the edge of chaos, its Lyapunov exponent being $\lambda = 0$. Under this point of view, the logistic map, equation 3.3 discussed in section 3.3, is not a complex system, in spite of presenting critical dynamical behaviour for some well chosen values of the parameter a. The many-units criterion is not fulfilled.

An interesting example of a real complex system is the social behaviour of singing birds — see KREBS, ASHCROFT and WEBER (1978). A bird acquires some advantages if it is able to imitate the song of another bird, for instance being allowed to obtain food inside the other's territory. In this way, evolution by selection pushes the singing ability to be "complex" (in the colloquial sense), in order to avoid imitations from the point of view of the territory owner. From the point of view of the intruder, the imitation ability, and thus its own singing ability is also pushed to be more "complex", as in the red queen effect — see, for instance, KAUFFMAN (1993) or BAK (1997). It is not an easy task to measure this singing "complexity". What one does is trying to recognize songs as composed by combinations of other simpler songs, by analysing audible records. The larger the repertoire of a singing bird, the more "complex" it is considered. This kind of analysis, however, is very imprecise, and the results are very unsatisfactory under a quantitative point of view. The first question a physicist poses in such a situation is: where are the power laws?

This is one of the best prototype situations where computer simulations are supposed to have a fundamental contribution: they can help to deeply understand the problem under study. In a seminal paper, SUZUKI and KANEKO (1994) (see also PAPA and TSALLIS 1998) studied this problem, replacing birds by logistic maps . They took a population of many birds, each one represented by a logistic map, equation 3.3, with its characteristic control parameter a. At the beginning the distinct values of a for each bird are tossed at random in the interval $0 < a < 4$. The dedicated reader who has performed the numerical experiment proposed in section 3.3 knows that $a > 4$ is uninteresting. She/he knows also that a logistic map presents a very simple, not "complex" behaviour for most values of a. For $0 < a < 1$, for instance, the logistic map always ends at $x = 0$ after some few iterations. For $1 < a < 3$, a similar behaviour is found: it always ends at the same point $x = 1 - 1/a$. For $3 < a < 3.449$, the behaviour seems to be a little bit more "complex", the final situation being an alternance between two well defined values of x, as the dedicated reader has observed by comparing x_t with x_{t-2}. She/he certainly agrees that this is again the same boring, repetitive destiny, now with a doubled period. For $3.449 < a < 3.544$, by comparing x_t with x_{t-4}, one can observe again the same repetitive behaviour, where the period has suffered a further doubling. All these cases cannot be considered complex (no longer the colloquial sense). On the contrary, they are simple, in the sense that a simple, periodic final situation arrives after a **fast** transient regime. This transient is fast due to the exponential decay corresponding to $\lambda < 0$.

The non-trivial, non-repetitive behaviours occur only for the critical values of $a = 1$, 3, 3.449, 3.544 etc. All these cases correspond to $\lambda = 0$. The reader can argue the final destiny observed on her/his computer is also repetitive, although the machine spends more, much more time to reach this final situation (provided the reader, besides being dedicated, has also patience enough to wait for the convergence). OK, but only the limited machine precision, not the map itself, must be blamed. Would the computations be performed on an ideal, infinite precision machine, as already analysed in section 3.3, then the value of x would never converge in those critical cases: they correspond to what we can call **infinite** transients. Also for $3.570 < a < 4$, where chaotic behaviour is found, there are no repetitions at all (except for some pathological cases where odd periodic behaviours can be found). A chaotic attractor is not a simple sequence of P particular values of x repeated again and again after each period P. On the contrary, a value of x is never repeated (on the ideal, infinite precision machine) within a chaotic attractor. Nevertheless, the chaotic regime also presents a **finite** transient, the attractor being reached exponentially fast. In this sense, the chaotic regimes observed for $3.570 < a < 4$ are indeed not complex, not critical: they are as simple as the periodic ones, concerning how fast the final situation is reached.

As mentioned above, a logistic map is not a complex system. Nevertheless, a set of various logistic maps can be, provided they exert mutual conflicting influences on each other and evolve dynamically according to these influences. The dynamical rule defined by Suzuki and Kaneko, inspired by the social behaviour observed in birds, provided such conditions. As already mentioned, they took a set of $B \sim 10^3$ birds, each one represented by a logistic map with its characteristic control parameter a. At the beginning, a random value of $a(i)$ ($0 < a(i) < 4$) is assigned to each bird i. During the whole process, one keeps in memory for each bird the constant $a(i)$ as well as the current value $x(i)$ of its dynamical variable (every time bird i iterates once more its map, this value is updated). The dynamics is based on two-player imitation games. Suppose bird i will imitate bird j. First, their maps are iterated T_0 times, in order to eliminate possible fast transients. After that, imitator bird i enters into a calibrating stage by iterating T_{cal} times its own map $x(i) = a(i)x(1-x)$, where $x = \epsilon x(i) + (1-\epsilon)x(j)$, and ϵ is a small parameter coupling birds i and j at this stage (at the same time, in parallel, bird j also iterates its map T_{cal} times, but with no coupling at all). At the end of this calibration stage, the current values are $x_0(i)$ and $x_0(j)$ for the dynamical variables corresponding to birds i and j, respectively. Now, bird i decouples from j and delivers its own song which is the sequence $x_1(i), x_2(i) \ldots x_T(i)$, obtained by T further successive iterations of its map, starting from $x_0(i)$. Analogously, bird j delivers its own song $x_1(j), x_2(j) \ldots x_T(j)$, starting from $x_0(j)$, to be imitated by i.

The quality of the imitation gives a score $D(i,j)$ to bird i. Different rules are tested by Suzuki and Kaneko, in order to compute this score. Let's consider only one, the squared Euclidean distance $D(i,j) = \sum_{t=1}^{T}[x_t(i) - x_t(j)]^2$ between the two songs. Then, the whole process is repeated again, by exchanging the role of the two players, and another score $D(j,i)$ is assigned to bird j. The winner is the best imitator among birds i and j, corresponding to the smallest score. The looser is elimitated from the population, replaced by a new offspring from the winner. This offspring is simply another logistic map with control parameter a slightly different from that of the parent bird, i.e. $a(\text{offspring}) = a(\text{parent}) + \eta$, where η is a small number randomly chosen in the interval $-W < \eta < W$. The parameter W corresponds to the mutation rate. Another pair of birds i and j is then randomly chosen, and the whole process repeated again, and so on. After B such repetitions (equal to the total number of birds), one says the simulation process has completed one step.

Many versions of this model were tested. For instance, the birds can live on a square lattice, each one playing only against its four nearest neighbours. Alternatively, any pair i and j of birds can be randomly chosen to play one against the other, with no neighbouring territories. The replacement of looser individuals by a winner's offspring can be done only after each bird has performed many imitation games against others, computing the average score of each one. The dependence of the system on its various numerical parameters $(B, T_0, T_{\text{cal}}, T, W)$ was also tested. The final situation, after many simulational steps, however, presents always the same behaviour: the distribution of a values among the final population of birds presents pronounced peaks around the known critical values of a for the logistic map. This dynamical system presents, thus, self-organized criticality. Also, one can compare the averaged scores of birds, during the whole simulation, and the result is a clear advantage for birds with critical a values. Critical, complex sequences of x are hard to imitate. Also, a critical sequence is more likely to be a good imitation of some given pattern. This system dynamically drives itsef to the edge of chaos like in the new theories about the evolution of species as reviewed by KAUFFMAN (1991), (1993) and (1995). What is new in these theories?

As Charles Darwin has already taught us a century and half ago, the optimization process followed by natural selection is dynamically performed through **small** steps, in order to keep always the advantages already obtained until the current state. Large steps towards better performances are normally very dangerous because the blind walk through unknown, new options can lead to evolutionary dangling ends. In such a situation, one must to keep active the possibility of returning back, trying to follow another path. A fast evolution through large steps forbids this possibility, because the situation some few steps ago may be so different than the current one that it is perhaps already extinct. Translated to our modern mathematical language, this means Nature does not follow a chaotic ($\lambda > 0$) search for forms better than the current ones, on the contrary, this search corresponds to a non-chaotic dynamics. On the other hand, it is very important to have access to new options very different from the current situation, allowing to evolve to better forms and performances. This is not the case observed in predictable systems ($\lambda < 0$), where the destiny of the dynamical evolution is the boring, repetitive behaviour of visiting again and again the same finite set of states. The advantage of evolving first to, and then **on** the edge of chaos ($\lambda = 0$) is just the large spectrum of available options, compared with predictable systems, and the safety provided by the long term memory, compared to chaotic systems (see also section 3.1).

The computer-implemented imitation games performed by Suzuki and Kaneko beautifully exibit all this behaviour proposed by the quoted new evolutionary theories. In this case, due to the simplicity of the individuals (the logistic maps), the result they obtained becomes very clear, perfectly understandable by our intuition. Indeed, following our analysis of section 3.3 concerning infinite transients, it is very plausible that critical sequences are more adequate than non critical ones. Other similar models can be implemented on computers. MARTINS, PENNA and DE OLIVEIRA (1999) have made such a study. The situation is the same that as treated by Suzuki and Kaneko, now with neural networks instead of logistic maps playing the role of individuals. A neural network is itself a complex dynamical system, with many units evolving under conflicting internal interactions. Thus, we are now testing the imitation game idea for individuals which are themselves complex, besides the complexity arising from their conflicting mutual influences. Let's first introduce the main features of a neural network, needed to understand the imitation games played by a population of them. First, we will change the name to **recognition games**, for reasons the reader will understand soon.

A neural network is an artificial device, say a computer program, designed to recognize patterns. Those patterns are normally coded as a sequence of bits whose values are 1 or 0. A black and white photo, for instance, may be coded in this way simply by considering it as a large square lattice with only black and white pixels (pixels are elementary small cells of the square lattice). This is indeed the code used in order to transmit a photo from a far place to the office of a newspaper which will publish it tomorrow morning. Also colour photos can be coded in the same way, for instance by using 8 bits per pixel, instead of only one, allowing for 256 different colours. Written texts can also be coded in the same way, using 8 bits per letter: 256 different symbols are enough to code capital and lower case letters, digits, punctuations, accents, etc. Independent of the application, we will consider a pattern as a sequence of N bits 1 or 0. A neural network has a set of P fixed patterns stored in its memory. It works by recognizing (or not) an extra pattern presented by the user, i.e. it chooses one among its own P stored patterns, namely the particular one which fits best the currently presented one. This is not an easy task, because some criterion of similarity must be adopted. Physicists have entered into the research about neural networks since the pioneering work of HOPFIELD (1982). In his model, the criterion of similarity is automatically included through an energy minimization process.

Neurons in our brain also code informations in this binary way: each neuron can be found in only two states, quiet or firing, at a certain instant of time, and can flip from one state to the other according to the influences received from other neurons. Those influences are conflicting, and the whole system dynamically evolves according to them. Of course, our brain can be classified as a complex system (indeed, it is considered as the prototype complex system). Some neurons receive inputs directly from the environment, as the light detectors present in our eyes, allowing the organism as a whole to react to external inputs. Here, we are not interested in the chemical/electric transmission mechanisms of these influences between neurons, occuring in a microscopic scale, but only in the macroscopic effect: a neuron receives inputs from others, some of them driving it to fire and others to stay quiet. This neuron, thus, adopts a compromise solution following the stronger sum of these conflictive influences. Many other biological ingredients (as noise, latent times, and others) make this system much more complicated than the simplified description above. Following once more the tradition of physicists, we will ignore all these complications.

The Hopfield model for neural networks mimicks the simplified view described in the last paragraph. Each neuron is represented by a bit 0 or 1 (quiet or firing). Neurons i and j interact with each other through the synaptic strength J_{ij}, a number which if positive, means that i and j present the tendency to stay in the same state (both quiet or both firing). On the other hand, for a negative coupling J_{ij}, neurons i and j tend to stay in opposite states (one quiet, the other firing). The distribution of these J_{ij} values, the synaptic matrix, must be previously designed, as we will discuss later, and will be kept fixed. Each neuron i is linked to many others $j = 1, 2, 3 \ldots$, suffering thus conflicting influences from all of them. In order to decide whether neuron i must be flipped or not, one adds together the strengths of all these influences, some positive and others negative, following the sign of the sum. The same procedure is repeated to update another neuron i', then another i'', and so on. After a few such updates, all neurons get into a stable situation, an attractor of this dynamics, and this is considered as the "recognized" pattern. However, starting the whole process again from another initial distribution of 0- and 1-bits for the N neurons, the "recognized" pattern may be a different one, another attractor.

The whole set of attractors corresponds to the network total memory. They represent all possible recognized patterns, each attractor being one of the P memories (each attractor will be called one memory). They depend on the values of J_{ij} actually implemented, and thus these values must be carefully designed in order to match each memory to one of the patterns the user wants to store in the network. This J_{ij} design is done before starting to use the network. Suppose one needs to construct a neural network to recognize P criminals whose photos are in the police album. Each photo corresponds to a sequence of N 0- or 1-bits. First, one needs to determine the N^2 values of J_{ij} in such a way that each photo is one attractor, with no other (spurious) attractors. There are many recipes to perform this task, the first one (the Hebb rule) proposed by Hopfield himself. Starting with $J_{ij} = 0$ (tabula rasa), one adds $+1$ for each memory (photo) where bits i and j are equal, and -1 for each other memory where bits i and j differ from each other. So, J_{ij} is the sum of P terms ± 1, and may be positive or negative. This is not an efficient recipe, but it works provided the memories are decorrelated (one cannot have similar photos), and the number P of memories must be small.

For a large enough N, this system presents two distinct regimes. First, for

$P < 0.14N$, the recognition ability of the network is good: given some input, say, an extra picture of criminal X (not the one found in the police album), then the above described dynamics drives the network to the photo of X in the album. Among the P memorized patterns available, the network automatically chooses the one most "similar" to the input. Second, if $P > 0.14N$, the network looses its recognition ability, i.e. spurious attractors appear, the memories themselves may no longer be attractors, etc. If the police album is too large, one can give a picture of one criminal as input and retrieve an unrecognizable, blurred superposition of many others, after applying the dynamics. The transition point $P = 0.14N$ is like the critical points $a = 1, 3, 3.449, 3.544$ of the logistic map: each of them also divides the system in two regimes (for instance, one has a single fixed point for $a < 3$, whereas a two-point attractor appears for $a > 3$). The neural network, however, presents only one such a transition point, whereas many of them appear in the logistic map. There is also another technical difference we will quote here only to satisfy theoretical physicists who are normally very proud of their analytical results: the neural network transition is a first order one. This means the transition point is not strictly critical. Nevertheless, the range of correlations is maximum at the transition point, although finite, and this is enough for our purposes.

Neural networks are a fascinating subject, and much more than the simplified explanation above is already known. Many different models, theoretical approaches, experimental data and practical implementations are available in the huge literature concerning this subject. This literature is spread over journals of biology, physics, informatics, etc. We are not able even to give a small list of fundamental references covering the subject. Would we try to do that, and the resulting list would be like a fractal or a Swiss cheese, with more holes than material. Instead, we direct the reader to only one reference — DE OLIVEIRA (1997). Doing so, instead of blaming any of the authors of the present book, the reader can blame the author of this reference for the many lacking informations.

The model developed by Solange Martins in Niterói considers a population of B neural networks (birds), all of them with N neurons. Bird i has P_i randomly chosen memories, and the values of P_i for each bird are also randomly chosen. At the beginning, some birds are overloaded ($P_i > 0.14N$), while others are below the maximum recognition capacity ($P_i < 0.14N$), according to flat distribution. They play the recognition game in pairs: bird j passes one of its own memories, say J, as an input to bird i. The latter updates this configuration dynamically, according to its own synapses, until convergence to some attractor, say I. Then, one measures the Euclidean distance between I and J, this result being the recognition score assigned to bird i. Then the reverse is performed, and another recognition score is assigned to bird j. The looser is eliminated, replaced by one winner's offspring. This offspring is a copy of the parent with mutations, i.e. it has the same memories except for one missing ($P(\text{offspring}) = P(\text{parent}) - 1$), or one random extra memory ($P(\text{offspring}) = P(\text{parent}) + 1$). After many repeated steps of this dynamics, the distribution of Ps is peaked near the transition point, just as already found in the imitation games. The difference here, as already commented, is the complexity of each individual itself. Although birds in the final population have approximately the same value of P for all, near the transition point, these birds are **different** from each other. A logistic map is characterized only by its parameter a, thus two final birds with the same a are identical within the imitation games of Suzuki and Kaneko. On the other hand, two neural network birds can have the same number P of memories, but the memories themselves are not the same: they are different birds, allowing a much richer structure. Various interesting quantities can be measured, as, for instance, the population diversity. Work along this line is in progress.

4 Microscopic Market Simulations

4.1 Stock Market Model of Levy, Levy and Solomon

The mathematical description of economic markets has a history of one century (PARETO 1897, BACHELIER 1900, MANDELBROT 1966/67, 1982, 1997), and economists like ZHANG (1991) applied physics concepts to it. If the present authors would have a reliable theory predicting the stock and currency markets of next week, they would be too busy spending their profits to have time for writing this book. We would already be happy if we could merely understand economics, or if at least others like our governments would. Why is unemployment in the country of the senile author today one order of magnitude above that in his youth ? How was the formerly 80 percent monthly inflation in the country of the two other authors now contained around one percent ? How do fluctuations in the stock market arise ? Already a decade ago, the Santa Fe Institute published proceedings on the economy as an evolving complex system; among the editors are a physics Nobel laureate and an economics Nobel laureate (ANDERSON, ARROW and PINES 1988).

Apparently these fluctuations are not well described by a simple random walk which would lead to a Gaussian distribution of the changes from one trading day to the next. (For times beyond one month the stock market fluctuations might become Gaussian.) Paul Lévy invented distributions (MANDELBROT 1966/67) which seem to fit better (see TSALLIS 1997 for a recent review of such distributions), and truncated Lévy distributions, fractional Brownian motion in multifractal time (FISHER et al 1997) or other modifications (GOPIKRISHNAN et al 1998) approximate the tails even more accurately (see MANTEGNA and STANLEY 1997 for a recent review). Several books on these phenomenological descriptions seem to be just out or in preparation (e.g. BOUCHAUD and POTTERS 1997, MANDELBROT 1997, see also KERTESZ and KONDOR 1998). However, in the tradition of statistical physics, we want to get such results out of microscopic models dealing with single investors and their more or less rational decisions, as we study Ising models simulating single spin flips. David Mermin in Physics Today of March 1997 remarked on the speech of the 1996 Nobel prize winner in economics: It's like physics, except that physics works. Well, we know many physics phenomena which we do not understand; let us see if we understand some economics through computer simulations of individual agents. (For early attempts see FOLLMER 1974; CALLEN and SHAPERO 1974; TOULOUSE and BOK 1978.) A general philosophy of econophysics was outlined by AUSLOOS (1998).

The LLS model of LEVY, LEVY and SOLOMON (1994/95) seems to be that microscopic model about which the largest number of papers are published at present. The authors assume that investors on the stock market want to maximize their relative profits. Thus a poor associate professor investing thousand dollars and a year later getting 2000 dollars back is very happy; a rich full professor getting 1.002 million dollars back from a million dollar deal may feel driven to drink. Each investor remembers the past investments and believes that this past will be repeated. Thus new investment will be made where in the past the largest percentages in profits were reached. This investor behaves like the persistent walker already quoted in section 3.2.

Of course, if all investors have the same information and use it to take rational decisions, they will come to the same conclusion, value each type of investment equally, and no trade will happen. This would be bad news for employees in Wall Street, though such equal interests might be very helpful in international politics. In reality, markets function because investors (similar to physics researchers) believe they are better than their competitors and can make profit by judging the future more reliably. In the trading between major currencies, only a small percentage is necessary to pay for exported goods or imported services; most of the trade is speculative. Thus, the LLS model assumes that investors have a different memory span over which they average to guess the future performance of stocks. And the decisions are not completely rational but include some emotional component simulated by some random numbers influencing the price. Risk aversion is not included explicitely.

The LLS model divides investment into bonds of a fixed interest rate, like government obligations, and stocks of variable profit, like shares of the IBM company. Thus the bonds are called riskless (none of these authors is German or Brazilian). The investors distribute their wealth among stocks and bonds, depending on their judgment of the future. The constant interest r of the bonds causes a general price increase. This inflation is realistic and is taken into account in the simulations of the LLS model but is not the cause of the short-term fluctuations of the stock prices. Thus we set $r = 0$ for simplicity here.

The dividend D_t paid by the company and the change of stock price P_t from one time $t-1$ to the next time t give the profit (or loss) of the investor, and the relative profit

$$H_t = (P_t - P_{t-1} + D_t)/P_t \tag{4.1}$$

has to be compared with the profit $r = 0$ from the bonds. Positive H_t for past times t will encourage investors to shift their wealth from bonds to stocks, while negative H_t have the opposite effect. In most of the past simulations, only one type of stocks was traded.

If a fraction X of the investor's wealth W is invested in stocks, and the remaining fraction $1-X$ in bonds, the profit is XWH_{t+1} if the new relative profit H_{t+1} per dollar would already be known. Averaging over the past k time steps (trading days) in the above mentioned sense of enjoying relative gains more than absolute gains, the investor ignores risks and tries to maximize the average over the logarithmic profits, called the expected utility U:

$$U = < \log[W(1+XH_j)] >_j \tag{4.2}$$

where the time average goes over the past k time steps $j = t-k+1, t-k+2, \dots, t$. Here H depends on the company whose stocks are traded while X and W are different for different investors. If the past brought only profits, all $H_j > 0$, then the largest possible X value $X = 1$ is preferred: Everybody wants to buy stocks. If stocks only produced losses in the past, all $H_j < 0$, then everybody wants to sell stocks, and $X = 0$ is preferred. In general, if none of these extremes is the optimum, the investor determines the desired fraction X by setting the derivative dU/dX to zero:

$$\sum_j H_j/(1+XH_j) = 0 \quad . \tag{4.3}$$

Then this investor wants to own XW/P shares of price p each, which may be more or less than the presently owned number and thus leads to the wish to buy or sell. This holds for all different investors, and the market price P is determined such that supply and demand are in equilibrium: The higher the price P is, the smaller is the number of shares XW/P this investor can buy.

To take into account some nonrationality of investor decisions, the optimal X fraction calculated above is changed by a small random amount, with a narrow Gaussian distribution. In the calculation of X no debts can be made, and some cash should remain available to buy a beer, i.e. $0 < a < X < b < 1$, where a is a constant close to 0, and b is a constant very close to 1.

For more details and justification we refer to LEVY, LEVY and SOLOMON (1994, 1995) or to the book they are writing for World Scientific (Singapore). Now we turn to the results of this model.

After some equilibration process, the stationary distribution obtained from computer simulations seems independent of reasonable initial conditions. If all investors have the same memory span k and decide completely rationally, there is no exchange market at all. With moderate noise there are enormous roughly periodic jumps of the prices up and down, separated by regions of nearly constant stock values. Such behaviour is understandable as long as eq.(4.3) leads to $X = 0$ and 1 (or more precisely, $= a$ and b) for all rationally deciding investors. The curves look better if the investors form groups with different memory spans or have randomly distributed memory spans k. Particularly when the largest and the smallest memory span are drastically different and several choices of k are mixed together, the market values fluctuate in a realistic way similar to true markets (LLS 1995, fig.14; HELLTHALER 1995, fig.4). In particular, the distribution of the changes from one time step to the next deviates clearly from a Gaussian (LEVY and SOLOMON 1996) and is compatible with a Lévy distribution (MANDELBROT 1967) and actually observed distributions.

Lévy distributions of index ψ decay in their tails with a power law of exponent $1 + \psi$ instead of exponentially. Thus higher moments of the distribution diverge; see TSALLIS 1997 for an introduction. The simulations of the Jerusalem group (LEVY and SOLOMON 1996) do not seem accurate enough to check for truncation effects (MANTEGNA and STANLEY 1997) in the Lévy distributions. Note that these distributions were invented by Paul Lévy decades ago, not by Moshe or Haim Levy of LLS in the 1990's.

However, as is common in physics, HELLTHALER (1995) went to the "thermodynamic limit" by increasing the number of investors from 100, the typical value of LLS (1995), to several millions, keeping the other parameters constant (and in particular simulating only one company stock). What looked like a reasonable market price fluctuation for a few hundred investors became according to Hellthaler a smooth and roughly periodic function of time for 10^6 players. (See STEIGLITZ et al 1996 for periodicities in another microscopic market model.) Thus the LLS stock market model with many investors and one stock no longer describes the unpredictable fluctuations of the stock market. There may be two explanations: First, one has to increase the number of companies whose stocks are traded proportionally to the number of investors in order to get good results; this was tried by KOHL (1997) (who again is not the one ruling Germany 1982-98) but with only limited success thus far. Second, the present unprecticable markets would indeed become more predictable if more people would invest; present markets might be dominated by a few professional investors working with big funds, so that the numerous private investors do not play a crucial role in smoothening out fluctuations. Also in the latter case, however, the LLS model would no longer be realistic since after some time the quasi-periodic oscillations of the prices of the Hellthaler simulations would be noted by all investors such that these anticipate them and change their behaviour accordingly. (See below for the learning traders of Arthur et al and Palmer et al.; in a different model with many stocks but only one investor, MARSILI, MASLOV and ZHANG (1998) found a connection with the Kardar-Parisi-Zhang equation of surface growth, in infinite dimensions.)

Also the distribution of wealth is of interest: How many investors become rich and how many poor, if initially everybody had the same wealth? For this question, power laws seem to be valid: Very few are rich, and much more are poor. In the USA, one percent of the population holds 40 percent of the wealth (LEVY and SOLOMON 1997). Similar power laws, known from the scaling theory of critical phenomena, occur in other aspects of self-organized criticality (BAK, TANG and WIESENFELD 1987, BAK 1997), like the Gutenberg law for earthquakes on the Richter scale or the Zipf law for the frequency in which words occur in natural languages. The PARETO (1897) law for income, applied instead to wealth, asserts that the probability distribution $P(W)$ for the wealth of individuals obeys a power law

$$P(W) \propto 1/W^{1+\psi} \tag{4.4}$$

with some exponent ψ typically between 1 and 2. Using the Forbes 1996 list of the richest 400 Americans, LEVY and SOLOMON (1997) found ψ to be slightly below 1.4.

These same authors, with the crucial help of Givat Ram (LEVY, SOLOMON and RAM 1996), gave a simple explanation of such power laws, which was already hinted at by SIMON (1957) for city growth (with an acknowledgement to Mandelbrot): Because of the logarithmic definition of the utility U (based on the relative and not the absolute profits, eq.(4.2)) the more an investor owns the more will he or she invest. Thus the wealth W follows a random walk on a logarithmic scale, that means $\log W$ is random: "Power laws are logarithmic Boltzmann laws" (LEVY and SOLOMON 1996). To get the correct exponents, however, some degree of socialism is needed (LEVY and SOLOMON 1996): Bankrupt investors are supported by welfare. Thus the random walk in $\log W$ is bounded from below. Only then can the resulting power law exponent be fitted to $\psi \simeq 1.4$, appearing both as the exponent of the wealth distribution and as the index of the Lévy distribution. And this $\psi = 1.4$ was also found empirically in the Standards and Poor index of the Wall Street stock market (MANTEGNA and STANLEY 1997). SORNETTE and CONT (1997) have analyzed mathematically the properties of this diffusion in $\log W$ space with a repelling boundary to the left.

4.2 Other Market Models

Earlier than any of the econophysics papers here were the economists KIM and MARKOWITZ (1989) who simulated two types of pension fund strategies: Rebalancers sell stock if they own appreciably more stock than cash, and buy stock if they own appreciably more cash than stock. Portfolio insurers, on the other hand, follow a Constant Proportion Portfolio Insurance rule such that the portfolio cannot lose more than, say, 25 percent of assets during the life of the insurance plan (at most 3 months). Each pension fund has deposits and withdrawals of customers at random times and in random amounts. The resulting market becomes increasingly wild with increasing ratio of portfolio insurers (or other trend followers) versus rebalancers.

An early simulation of TAKAYASU et al (1992) works with buyers and sellers who trade only if their desired price difference exceeds a minimum margin; in that case the highest bidder buys. After buying, the price for which the buyer is willing to buy more is reduced; the opposite change occurs after selling. Moreover, the desired prices change also randomly. The model is somewhat similar to particles in one dimension which attract each other if their distance exceeds a minimum distance (the above-mentioned margin). The resulting price fluctuations then were similar to a Lorentzian as a function of frequency. With some short-time memory included (the same for all traders), such that the desired price at time $t+1$ depends not only on that at time t but also on that at time $t-1$, a crash of the market price could be produced, and the power spectrum was proportional to $1/(\text{frequency})^2$ like a random walk. The final results depend on the initial conditions, with a zero Lyaponov exponent (as discussed for leaky faucets in section 3.3): Self-organized criticality; this market moves to the edge of chaos. This work was recently continued by SATO and TAKAYASU (1998) and by EBELING, SCHARNHORST and KARMESHU (1998), bridging the gap between the microscopic and the phenomenological approach. The latter paper also emphasized techniques similar to biological evolution (KAUFFMAN 1993, 1995) to investigate the dynamics of technological change: Hill climbing in a complex adaptive landscape.

BAK, PACZUSKI and SHUBIK (1997) (see also TANG and TIAN 1998) assume that each trader can either own one share of the stock or none; the money not invested in stocks brings a constant "riskless" interest as in the LLS model. Again, mostly only one company is traded. There are two types of traders: rational ones which look at the fundamental values of the company and try in different degrees to avoid risks; and noise observers who without much regard for the performance of the company mainly look at the market and the behaviour of other noise traders. The model also simulates the volatility of stock markets: A big change in one day is likely to cause bigger-than-usual changes during the following trading days. In the limit of only noise traders which moreover ignore each other's behaviours the model of BAK, PACZUSKI and SHUBIK (1997) becomes equivalent to a reaction-diffusion model of the type $A + B \to 0$, where particles and "antiparticles" annihilate each other. This model was already solved mathematically (BARKEMA, HOWARD and

CARDY 1996).

The more general cases, when noise traders imitate each other or are mixed with rational traders, are treated by computer simulation with $\sim 10^3$ investors. When noise traders dominate, the feared "bubbles" occur when the price rises far beyond what the rational traders regard as reasonable; then the noise traders are all by themselves, until the bubble bursts, the stock price crashes, and normal behaviour returns. The authors compare this with the depinning transition in solid state physics. Sequential and not simultaneous updating is used, and thus there is no price determined by a central authority. Perhaps, though the authors do not mention this, trading for houses or used cars is more appropriate for this model than the stock exchange in Wall Street.

The resulting fluctuations differ from a Gaussian distribution and thus agree with reality; the data were not accurate enough for a quantitative comparison with Lévy distributions. Surprisingly, some results are history dependent and thus influenced by the initial conditions. We refer to this paper for the many interesting variants discussed there. Moreover, the authors should be thanked for listing many relevant economics papers, while we mention here mainly those published by physicists in physics journals.

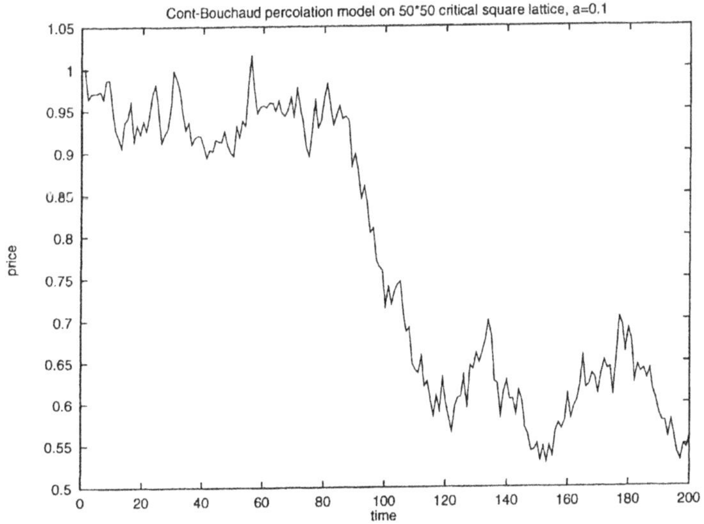

Fig.4.1 - Bad luck at the stock market: The Cont-Bouchaud herding model applied to the square lattice at its percolation threshold just gave in this small-scale simulation a rather catastrophic crash.

Also STEIGLITZ et al (1996) use this balance of rational traders and noise observers; in addition they have producers of varying skills which have the choice of farming for food or mining for gold; some of the speculators (noise traders) can be made "jumpy" and then react to much smaller profit margins than their colleagues. Without speculators the market price shows huge periodic oscillations; introducing the speculators after thousand iterations drastically dampens these oscillations and lets the gross domestic product grow faster.

The economist LUX (1997/8), too, distinguishes between rational and noise traders (fundamentalists and chartists). The latter ones can be optimists or pessimists; the probability to switch their mood depends exponentially on both the prevailing mood and the actual price trend. Also switches from a fundamentalist to a chartist attitude and back happen randomly, with probabilities depending exponentially on the difference of the current profits made by these groups. The simulations show clear correlations between the volatilities at consecutive iterations, while the sign of the price changes is not predictable, in agreement with reality.

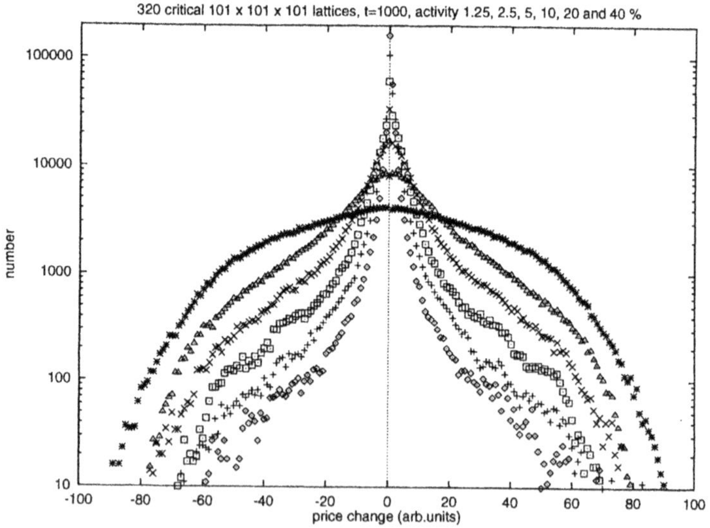

Fig.4.2 - The critical Cont-Bouchaud herding model shows in this histogram of simulated price changes a crossover from a power law to a Gaussian distribution, if for a fixed number of traders their activity increases from very small (diamonds) to very high (stars) values. From STAUFFER and PENNA 1998

A simpler model suitable for exact solution was suggested by CONT and BOUCHAUD (1998). Traders i either want to buy ($S_i = +1$), sell ($S_i = -1$), or not trade ($S_i = 0$). They form clusters of random infinite-range bond percolation (called "random graphs" in the mathematical literature); all traders within the same cluster have the same wish S_i. The price of the single stock traded changes proportional to $\sum_i S_i$. When at the percolation threshold an infinite cluster appears, it causes a crash or boom and signifies the limit of applicability of this model. It gives an exponentially truncated power-law distribution of large price fluctuations, and a crossover to a Gaussian distribution for low probabilities of not trading. Thus the model points to aggregative behaviour ("herd behaviour") of economic agents as a possible source of intermittency in financial time series. The generalization to nearest-neighbour connections on lattices in one to seven dimensions gave similar results (STAUFFER and PENNA 1998; STAUFFER et al. 1999; CHANG and STAUFFER 1998; see CHOWDHURY and STAUFFER 1999 and JOHANSEN, LEDOIT and SORNETTE 1999 for Ising-like models), as seen in Figs.4.1 and 4.2. The Cont-Bouchaud infinite-range case is very similar to the model of KAUFFMAN 1995 for the origin of life by auto-catalytic reactions. Another herding model was invented by ILINSKI and STEPANENKO (1998) who also compare speculators with movable electric charges and their screening effect: Any local opportunity for a large profit ("arbitrage") will be smoothened out by speculators moving into that market, just like ions in an aqueous solution cluster around any electric charge and thus screen the effect of this charge.

Also CALDARELLI, MARSILI and ZHANG (1997) (see also MARSILI and ZHANG, 1997) look at only one stock and its price fluctuations. But they apply Darwinian evolution to the population of traders: Each trader randomly selects a strategy of buying and selling at the beginning and gets the same capital; then at each time step that trader with the smallest remaining capital is eliminated and replaced by a new one with randomly selected strategy. The amount which trader i wishes to buy or sell is given partly by a nonlinear function

$$f = x/(1 + (x/2)^4); \quad x = \sum_{k=1}^{\ell} \eta_{i,k} < \partial^k \log p / \partial t^k > \tag{4.5}$$

which depends on the time averages of the derivatives of the past market price p, with typically 10 to 100 time steps averaged over. As in the LLS model, it is log p and not p itself which determines the traders's judgment. The ℓ coefficients $\eta_{i,k}$ for each trader i are chosen randomly. In addition the trading decision is influenced by the present amount of stocks and other wealth held by the investor to keep them in a reasonable equilibrium. Also some noise is introduced into the decisions. A common price p for all traders is found by time-averaging over past supply and demand:

$$p_{t+1}/p_t = \langle \text{demand} \rangle / \langle \text{supply} \rangle \quad . \tag{4.6}$$

The resulting curves $p(t)$ of price versus time look realistic, and the price fluctuations from one time step to the next deviate strongly from a Gaussian: Like Lévy distributions they have fat tails with a power law decay $\propto \text{change}^{-1-\psi}$ and ψ close to 1. (Politically correct papers would describe these tails not as fat but as gravitationally challenged: D.L. Hunter, priv. comm.) The wealth distribution of the traders also follows a power law (similar to LEVY and SOLOMON 1997), but with a different exponent. (The same Darwinian rule to throw out unsuccessful traders was applied by ISPOLATOV, KRAPISKY and REDNER (1998) in their different trading models; their mean-field and simulational approach gives wealth distributions following a Fermi function or a 1/wealth law.)

CHATAGNY and CHOPARD (1997) simulate the currency instead of the stock market. Investors base their profit expectations on the "momentum", the difference between the last known price and the weighted average over all past prices. Different dealers have different time horizons within which they must close their open positions, from one day to several months. The results look chaotic, and indeed slight changes in the parameters produce a completely different evolution of the market. Details are given for parallel efficiency on an IBM-SP2 computer.

Perhaps the most sophisticated model are the learning agents of PALMER et al. (1994) and ARTHUR et al. (1997), involving bit-strings, mutations, and sex similar to section 2.5, but also a sunspot variable. Instead of following outdated rational expectation theories, each agent follows its own set of 60 rules for what to do under certain conditions. These conditions are coded as strings of about 75 bits. Since sex on the trading floor of the stock exchange might be difficult, a simplified form called "genetic algorithm" is used instead, which still recombines the bit-strings. So the rules adapt, like living beings, to the environment and work better and better; the rules needed for success turn out to change in time. Crashes and bubbles without simple explanations are found, as in reality. The wealth distribution is unequal for intermediate times, but for very long times those who were wealthy could become poor again, and the other way around. Lévy distributions were not discussed.

The model of DE SOUZA (1996), see also TSALLIS, DE SOUZA AND CURADO (1995), for stock markets uses nonlinear dynamics in a coupled map lattice model of a few hundred elements. The probability $p_i(t)$ of trader i wishing to buy at time $t+1$ is the sum of three terms: an inertia proportional to α_i (which has nothing to do with the random walk α of the previous chapter) to stay with past behaviour; an imitation tendency propotional to β_i to follow the behaviour of the other traders; and an "expert" probability proportional to γ_i to do the opposite of what the others are doing:

$$p_i(t+1) = \alpha_i p_i(t) + \beta_i < p(t) > -\gamma_i p_i(t) < p(t) > \qquad (4.7)$$

with free parameters α_i, β_i, γ_i and the average $< p >= \sum_j' p_i/(N-1)$ defined by the summation over all $N-1$ other traders j except $j = i$. If p_i would become larger than one or smaller than zero it is set equal to one or zero, respectively.

In the special limit of $N \to \infty$ with $\alpha_i = \alpha$, $\beta_i = \beta$, $\gamma_i = \gamma$ the average probabilities obey the logistic equation

$$< p(t+1) >= (\alpha + \beta) < p(t) > -\gamma < p(t) >^2$$

which is well known to lead to chaotic behaviour if $\alpha + \beta$ are large enough. In general the coefficients like α_i are distributed narrowly for well informed investors, and widely for other traders, and one may regard them as fixed or as changing randomly at every iteration. Additional noise does not change much. Initial conditions are finally forgotten after sufficiently long time. Variation of N between 30 and 3000 had little influence.

The resulting plot of $< p(t) >$ showed strong fluctuations, but a certain periodicity in time and anticorrelations from t to $t+1$ might exist. The market may collapse if everybody wants to sell or everybody wants to buy. When eight different companies are traded on the stock exchange, instead of one, then after some time one is much more in demand than the seven others. This monopoly can be avoided by government intervention to help failing companies, taken into account by a term proportional to $1/ < p >$ in the generalization of eq.(4.5). In the latter case realistic looking curves for demand versus time were found, and also for prices versus time (with additional assumptions on the relation of price and demand) with 8 stocks traded. No comparison of the fluctuations with Gaussian or Lévy distributions is given in DE SOUZA (1996).

SORNETTE and JOHANSEN (1997) construct a hierarchical model of traders imitating each other. Two traders form one group, two groups form one supergroup, and so on until we reaches the level of large banks, countries, and currency blocks. This corresponds to a Bethe lattice (Cayley tree) of branching ratio two. Each member of a group, supergroup etc is influenced directly only by the other member and by orders from superiors. (Apparently these authors follow the ideology of oldfashioned Theocracy that France is influenced only by Germany and not by the 13 other countries in the European

Union.) Members need different times to reach a decision to buy stocks; if one member of a group decided to buy, it shortens the time of the other to reach the same decision. This crowd effect causes the positive feedback postulated e.g. by ARTHUR (1990) for the economy in general. But it also leads to crashes, like the "bubbles" of Bak et al. (Apart from the different times to reach a decision, all traders are alike.) Mathematically the crash is reached when infinitely many traders want to buy, somewhat similar to Bose-Einstein condensation or the behaviour of helium 4 cooled down to two degrees above absolute zero temperature.

More precisely, the renormalization group treatment by these authors of their discrete hierarchy leads to log-periodic oscillations with time t of the form

$$\text{const} + (t_c - t)^\alpha[\text{const}' + \cos(\omega \log(t_c - t) - \phi)]$$

which differ from both the usual power laws $\propto (t_c - t)^\alpha$ at critical points in physics and the widespread periodic oscillations $\propto \cos(\omega t - \phi)$. The October crashes in Wall Street of 1929, 1987 and 1997 are reasonably fitted to this log-periodic oscillation for the period of two years before the actual crash (see also SORNETTE, JOHANSEN and BOUCHAUD 1996 and VANDEWALLE et al 1997/8; they are special outliers, not just tails of the fluctuations, according to JOHANSEN and SORNETTE 1998). A more sophisticated treatment leads to a more complicated formula which fits reasonably the seven years before the 1929 and 1987 crashes. The data noise is still of the same order of magnitude as the oscillations described by this theory, but fits of similar quality were made for earthquakes, fracture of materials, and other models, as cited by these authors (see SORNETTE 1998 for a thorough review). All together they make these log-periodic precursors of the big event at $t = t_c$ plausible. Needless to say it would be of great practical importance to predict the time of a big earthquake or stock market crash by analysis of small oscillations in suitable indicators.

Similar to thermal physics and to the political models of the next chapter, WEISBUCH, KIRMAN and HERREINER (1998) treat the relations between buyers and sellers on the Marseille fish market. At low "temperatures", an ordered situation arises where buyers stay loyal to their usual sellers, apart from occasional infidelities. On the high-temperature side of the phase transitions, the buyers shop around and are divorced from their sellers. Similar to the LLS model, the traders keep in memory their past experiences (profits or losses) and let them guide their new decisions.

Finally we mention that WEISBUCH et al (1997) combined economic modelling with the extinction of fish species by overfishing, mentioned in Chapter 2. They point out an inertia effect: If new fishing boats are bought, the owners want to use them some years before they may adjust to reduced fish swarms. Also, VANDEWALLE and AUSLOOS (1997/8) found similarity between the DNA sequences discussed in the preceding chapter and financial data, while the model of VANDEWALLE et al (1997/8) was reported by DUPUIS (18. September 1997) to predict a crash in the Dow Jones Index before the end of November 1997; it happened on October 27 of 1997 and gave huge profits to some econophysicists (Sornette, as cited in STAUFFER and SORNETTE 1998). And BULDYREV et al (1997) simulated company growth.

Does it make sense if physicists study economic problems using their own methods? Opinions seem divided. Economist Brian ARTHUR (1990) mentions the Ising model of ferromagnetism and what physicists call broken symmetry to explain how more than one economic equilibrium is possible and one of them is selected more or less randomly, e.g. for videotape standards (VHS versus Betamax). On page 27 in ANDERSON, ARROW and PINES (1988) he complains that in economics little work has yet been done on spatial self-reinforcing mechanics like this Ising model; and on page 251 there he is reported to believe there are techniques in chemistry, physics, and biology that may be used by economists. But can we trust someone who puts down FORTRAN as the 1950's-vintage computer language?

Physicist Sornette, on the other hand (while authoring "Crashes at Critical Points" using physics methods: SORNETTE and JOHANSEN 1997; JOHANSEN, LEDOIT and SORNETTE 1999), thinks physicist should better follow the economics literature (private communication in a critique of a draft of this chapter: ARTHUR 1994; BARRETT, GEWEKE, and SHELL 1989; FAMA and FRENCH, 1996; GENNOTTE and LELAND, 1990; HIRABAYASHI et al. 1993; ORLEAN 1991; PERAZZO et al. 1995; ROEHNER 1989; SHILLER 1989; VARIAN 1989). He thinks one cannot capture the behaviour of traders by simple rules which do not evolve; traders learn. And if enough traders believe markets will go up, then indeed they go up. Many models will give the correct power law distributions for price variations. It is the several-point statistics that play an important role on the market, as in e.g. ARNEODO, MUZY and SORNETTE (1998). Future research should also include risk aversion and stability, i.e. what should be the rules that the central banks should impose on the market to make it more stable.

Perhaps the situation is similar to biological ageing: There are many theories, physicist tend to follow those paths corresponding to their physics methods, and it is not yet clear which theory is best. More precisely, we may compare this econophysics situation with that of ageing in 1994 (chapter 1), before the Penna model of chapter 2 started to dominate the Monte Carlo simulations of ageing. For in economics now, there is no consensus among physicists which type of model is best.

Finally, economics Nobel laureate Markowitz commented on a draft of this chapter: "I believe that microscopic market simulations have an important role to play in economics and finance. If it takes people from outside economics and finance – perhaps physicists – to demonstrate this role, it won't be for the first time that outsiders have made substantial contributions to these fields."

5 Political Theories of Galam and Others

5.1 Social models

How come we always loose elections ? Why does a women scientist become an Iron Lady? Why are some people members of the academy, or have a high count in the Science Citation Index? Of course, the elites of all times always find nice reasons for their status, and indeed we are authors by the grace of God. However, randomness as described by Ising model simulations and other Monte Carlo approaches gives other reasons, and such approaches will be described here. (Again, we single out here a few papers from statistical physicists; they again cite sociological literature. A selection of reviews on self-organization in socioeconomic problems is published in SCHWEITZER 1997.)

Does this make any sense ? The human brain with 10^{15} synapses is more complicated than anything thus far simulated on a computer (up to 10^{12} Ising spins only: LINKE et al 1995; four times more percolation sites: MACLEOD and JAN 1998), and on our Earth live nearly 10^{10} people. Is there any hope for a theory of interactions in human societies ? Galam thinks yes, because of the principle of universality. A single atom can be quite complicated, with its many electrons, protons and neutrons, which in turn may consist of gluons, quarks and antiquarks, and perhaps subquarks, too. Nevertheless, the evaporation of a liquid, or the ordering of a ferromagnet, can be described reasonably well by approximating these complicated atoms by one-bit variables, like occupied or empty for fluids, and up or down for magnetic spins. The collective behaviour of these atoms ignores many microscopic details of the atoms as if it averages over all of them. Can mass psychology, group behaviour and election results be described by the same ideas of universality, ignoring many aspects of reality and looking only at collective effects similar to symmetry breaking in statistical physics (ARTHUR 1990)?

A successful example of this type of ignoring many aspects of reality are the nearly 400 year old Kepler laws. They assume that the planets are points running around the sun. Of course, the Earth is not a point, religious wars tore apart Kepler's Europe, and today traffic jams keep us away from where we want to drive to. Nevertheless for planetary motion the approximation of point masses was quite successful, even for planets not yet known to Kepler. Many aspects of real planets are barely relevant to the question how fast these planets rotate around the sun. In this spirit, BONABEAU, THERAULAZ, and DENEUBOURG (1995), KOHRING (1996), ANDREEV, BORODKIN and LEVANOVSKII (1996), GALAM (1997), and SUTTON et al (1998) look at human collective behaviour by ignoring many traits of the individual.

Two authors from a Department of History at Moscow State University and one from a Moscow Institute for Physics and Technology (ANDREEV et al 1996) simulated the strikes of industrial workers by four nonlinear differential equations for four different groups of workers with different degrees of involvement in strikes. Their data, analyzed by techniques known from chaos theory, agree qualitatively with the wave of strikes in Zsarist Russia about a century ago.

Now let us look at how group decisions and compromises might be modeled. GALAM (1990) warned about problems from majority voting as opposed to proportional voting in election districts. Let us assume r people vote for one of two parties in one election district; their representative, elected by the majority, votes on a higher level together with $r-1$ other representatives for one senator; r senators vote for one governor, and so on, with always r votes giving one election result ± 1. If the original voters have a probability p_0 to vote for the party denoted by $+1$, then their representative will vote $+1$ with probability $p_1 = R(p_0)$ and

$$R(p) = \sum_k B_{rk} p^k (1-p)^{r-k}$$

where $B_{rk} = r!/[k!(r-k)!]$ are the binomial coefficients, and the sum runs from $k = (r+1)/2$ to $k = r$ if r is odd. The election of senators then follows the same "renormalization group" transformation: $p_2 = R(p_1)$, that of governors follows $p_3 = R(p_2)$, and so on. After a few levels of this hierarchy, even if the original voters are nearly evenly divided and have only a slightly higher probability $p_0 = 0.51$ to vote for party $+1$, this marginal advantage is transformed into solid victories for the $+1$ party on all high levels, with the minority of 49 percent excluded from power. If r is an even number and a tie vote means that the ruling party stays in power, then this party remains in power even if the opposition has a sizeable majority among the voters. (However, this difference between even and odd r is relevant only for voting in small councils with all members always present and voting.) A lattice version of this problem was simulated more recently by GALAM et al. (1998).

BONABEAU, THERAULAZ, and DENEUBOURG (1995) explain the building of a pecking order as a result of randomness combined with high population density. Their individuals run around on a square lattice, and each time two of them meet there is a fight which one side wins and the other side loses. The probability for individual i to win against j is $1/[\exp(-\beta 2h_{ij}) + 1]$ where h_{ij} is the number of past victories of i, minus the number of past victories of j. Initially, nobody has any victories or losses, and this probability is $1/2$. After some time, the past is forgotten. This model thus is similar to the heat-bath updating algorithm in statistical physics, where $\beta = 1/k_BT$. It exhibits a sharp transition: At low densities the individuals still win about half their fights, whereas at high densities some win nearly all, and others lose nearly all. This transition might correspond in human history to the transition from nomadic or rural life to life in dense cities, thousands of years ago. Visa requirements for Christofer Columbus were less stringent than they are for the present authors. Similarly, locusts switch from solitarious to gregarious appearance and behaviour, when their density increases, as simulated by KIZAKI and KATORI 1998.

KOHRING (1996) simulates a modified Latane model (NOWAK, SZAMREIJ and LATANE 1990) concerning how opinions are formed in human society. Each invidividual i has two choices $S_i = \pm 1$, like Coke versus Pepsi. The influence of others is felt through

$$h_i = \sum_j [J^s_{ij}(1 + S_iS_j) - J^p_{ij}(1 - S_iS_j)] \tag{5.1}$$

where J^s is the supporting and J^p the persuasive force. People having the same opinion, $S_i = S_j$, may support each other in their choice, while people with different opinions, $S_i = -S_j$, may to try persuade each other to change their position. The summation excludes the case $j = i$. The opinion at the next time step is

$$S_i(t+1) = S_i(t)\,\mathrm{sign}\, h_i(t) \tag{5.2}$$

which is the no-noise (rational, zero temperature) limit of a probability proportional to $\exp(\beta h_i)$. The above interactions J are not constant in time but involve learning: successful persuaders get more persuasive powers, somewhat similar to the winners of Bonabeau et al in the preceeding paragraph. The model exhibits three possibilities: uniformity (ferromagnet) with most of the people having the same opinion; frozen (spin - glass like) with most of the opinion staying in the varying positions; and leadership, when one or a small group of individuals gains a strong influence on the opinion of the others.

GALAM (1997) basically looks at a random-field Ising ferromagnet with energy

$$E = -J\sum_{i,j} S_iS_j - \sum_i h_iS_i, \quad S_i = \pm 1 \quad . \tag{5.3}$$

Here each individual i has two choices, spin up ($S_i = 1$) or spin down ($S_i = -1$), which may correspond to voting in a two-party system; E is the amount of conflict which each of the individuals wants to minimize, J gives the strength with which individuals influence each other. The interaction term $-JS_iS_j$ gives an energy $-J$ if the two spins are parallel (bond is satisfied), and an energy $+J$ if the two spins are antiparallel (bond broken since $S_i = -S_j$). Thus $2J$ is the energy to break one bond, and on a lattice of N individuals, where each spin has q neighbours, the smallest possible energy (optimal solution) is $-qJN/2$ since each bond belongs to two spins. (Alternatively one can work with the variables $(1+S_i)/2$ which are zero or one; then it is natural to normalize the energy to zero if all these variables are zero.) In eq.(5.3), h_i is the individual preference given to one or the other choice. Galam favours conformity and takes $J > 0$, that means an individual i should follow the opinion of the neighbours j if there are no overriding personal reasons h_i to deviate. These neighbours may be all other individuals, or they may be only the nearest neighbours on a square or simple cubic lattice. The requirement to minimize E corresponds to zero temperature in physics or completely rational behaviour for humans; GALAM (1997) promises a follow-up paper including noise, that means non-rational decisions taken with probability $\exp(-\Delta E/k_BT)$ where ΔE is the amount of unnecessary conflict the individual is willing to create.

At finite temperatures, with small fields h_i equally distributed about zero, this random field Ising model has a spontaneous magnetization in three but not in two dimensions. The spontaneous magnetization is $\sum_i S_i$, and if it is non-zero, then one of the two choices $S_i = \pm 1$ is preferred by the majority of the population. If all local "magnetic fields" h_i vanish, we have the normal Ising model, and now eventually the whole population shares the same choice (nonzero spontaneous magnetization) in both two and three dimensions.

In passing we mention that economist CREMER (1997) suggested an even more complicated model to simulate the pricing strategy of multi-national companies, involving spin glasses, random fields, and more than two choices in a Heisenberg magnet . And HOFFMANN and PEKALSKI (1998) looked at various approaches, including the Penna model of chapter 2, to see how an immigrant population with initially higher birth rate evolves among the native population.

Biological evolution was simulated by SUTTON et al (1998) to find out how hunter-gatherer societies settled on their preferred mode of living by transmitting genes, with mutations, to their offspring. Since these authors neglected sexual genetic recombination, we explain their model through the teaching at universities, where asexual reproduction of research techniques from professors to graduate students is more common. Some professors are good in Fortran programming, some prefer C, some use Pascal, etc. They try to teach their techniques to their doctoral students, and try to produce successful papers. The students are bright and modify the research techniques of their professors, as for genetic mutations in biology. After each "generation", the research group is evaluated, and those with more papers in Nature get more money to hire new students. The simulations showed that after 500 generations, everybody followed the same strategy; but depending on the initial conditions, the surviving research techniques were different.

5.2 World War II

Now we discuss the theory of AXELROD and BENNETT (1993) and GALAM (1996) for World War II: How did the anti-Hitler coalition in Europe form? The authors collected the economic and political preferences of various countries and put them into something like an Edwards-Anderson spin glass with energy $E = -\sum_i h_i S_i$ and

$$h_i = \sum_j J_{ij} S_j, \quad S_j = \pm 1 \tag{5.4}$$

where each country S_i wants to align in the direction of the influence exerted by the others: $S_i = \text{sign}\, h_i$. (Physically speaking, the energy is made as strongly negative as possible.) Axelrod and Bennett find that their theory gives the correct alignments for nearly all countries; only Salazar's Portugal in reality behaved better than the model and stayed neutral while the model says it should have joined the fascist camp.

GALAM (1996), however, has criticized the method in which Axelrod and Bennett constructed their model. He writes that the Axelrod-Bennett model is not a really frustrated Edwards-Anderson spin glass, but rather a Mattis spin glass. Such spin glasses have $J_{ij} = \epsilon_i \epsilon_j J$ with a constant J and $\epsilon_i = \pm 1$. With the transformation $\sigma_i = \epsilon_i S_i$ the spin glass interaction then reduces to

$$h_i = J \sum_j \sigma_j, \quad E = -\sum_i h_i \sigma_i, \quad \sigma_i = \pm 1 \tag{5.5}$$

which is the good old Ising ferromagnet. Here, in the no-noise (rational decision, zero temperature) case discussed by Axelrod, Bennet and Galam, all σ_i have the same value, either $+1$ or -1. Basically, one then has put in what one wanted to get out, and the predictive power is not as large as hoped by Axelrod and Bennet: given any spin configuration S_i, there exists always a set of J_{ij} which will give that configuration as a unique minimum of the associated energy. Instead, GALAM (1996) makes the following alternative suggestion how to describe world politics.

This alternative assumes again an energy $-\sum h_i S_i$ with

$$h_i = \beta_i b_i + \sum_j (G_{ij} + \epsilon_i \epsilon_j J_{ij}) S_j \tag{5.6}$$

where $S_i = \pm 1$ describes the adherence of a country to one of two coalitions, like NATO and Warsaw Pact during the Cold War in Europe. The matrix G_{ij}

describes the traditional amount of cooperation (positive G) or conflict (negative G) between countries i and j, the ϵ variables are $+1$, -1 or 0 depending on the natural choice to be in one of the coalitions or to remain neutral, the J_{ij} are always positive and describe the advantages e.g. from sharing weapons within one alliance, also β_i is $+1$, -1 or 0, and b_i describes the advantages for a single country connected with joining the coalitions.

When $|G| \ll |J|$ and $\beta_i = \epsilon_i$, then the same transformation $\sigma_i = \epsilon_i S_i$ transforms this model into an Ising ferromagnet in an inhomogeneous external field b_i:

$$h_i = b_i - \sum_j J_{ij}\sigma_j$$

This model has for negligible b_i two possibilities of equal lowest energy: All $\sigma_i = 1$ and all $\sigma_i = -1$. (The b_i can favour one over the other, and so can the neglected G_{ij}). This case may correspond to the two camps of the cold war; the Berlin wall then was the Bloch wall (or liquid-vapour interface) between the two ferromagnetic domains. If on the other hand $J \ll |G|$ we recover the Edwards-Anderson spin glass in an inhomogeneous field. This model is known to have numerous local energy minima, and even without external fields can be solved exactly only by computer simulation (up to 1800×1800 on the square lattice; PALMER and ADLER 1998). This case may correspond to the fragmented Europe in earlier centuries and to the danger that the unification process for (Western) Europe comes to a standstill, a reversal and finally destruction. The present situation with the Warsaw Pact no longer existing but NATO expanding eastward means $|G| \ll |J|$ again with all the ϵ in NATO direction unchanged, but with the other ϵ changing. Policies which deviated from the search for the energy minimum are called "risky" by GALAM (1996) and correspond to positive temperature in physics.

Different computer simulations of RUCKAMP (1998) could not precisely reproduce the results of Axelrod and Bennett or the historical truth of World War II from reasonable input parameters, also not for the Galam model. These parameters are necessarily somewhat subjective, and world history might me more complex than an Ising model.

Rather generally, the rational (zero temperature) behaviour does not necessarily have to be the same as that for small but positive noise ($T > 0$). For example, in the Ising ferromagnet at all small but positive temperatures, the system orders into one or two big domains if we start with a random distribution of spins. At zero temperature (flipping spins randomly if this does not change the energy, flipping them with certainty if this lowers the energy, and flipping them never if the flip would increase the energy) the behaviour is quite different, and in higher dimensions many separate domains may coexist since their growth is stopped (DERRIDA, BRAY and GODRECHE, 1994; see CUEILLE and SIRE 1997 for further literature). This spin problem is the ageing of the Redner group (FRACHEBOURG, KRAPIVSKY and REDNER 1997) but has nothing to do with the Penna model for biological ageing.

Actually, the war model of ISPOLATOV, KRAPIVKSY, and REDNER (1996) refers to these coarsening simulations of Derrida et al since their simulations, mean field approximations and exact solutions are similar to the growth of domains in alloys and magnets: If two civilizations meet, the smaller one is annihilated completeley while the larger one is merely reduced by an amount proportional to the size of the smaller civilization. Perhaps this model is more appropriate to the colonization of the Americas five centuries ago than to conventional wars. Unfortunately, the model is one-dimensional, and only at the end a two-dimensional generalization is suggested.

For finite "temperatures", CHAKRABARTI and MARJIT (1995) look an a spin-glass like model of employment in economics, and also mention ageing within Game-of-Life cellular automata. Finally, GLANCE and HUBERMAN (1994) summarize simulations not only of the political changes of 1989 in East Germany and elsewhere, but also of the less important but more recurring question on whether to order cheap or expensive meals if the bill is shared by all diners.

6 Summary

Does this whole approach make any sense ? Certainly, we have cured neither ageing nor AIDS, the ecomomics modellers have not become rich by their simulations, we did not win any recent elections, and wars still exist. However, the experts from these fields outside physics have not necessarily been more successful according to these criteria.

This book, similar in spirit to those of KAUFFMAN 1991/3/5, summarized selected examples of recent research by theoretical physicists, using methods of statistical physics and its simulation. The results were published mostly in physics journals, but partly also in journals of the fields to which the methods were applied. This research is thus open to productive criticism by the non-physics expert, and we hope this book will encourage and facilitate such criticism. We also hope to encourage other physicists to try similar interdisciplinary work using their own methods, either in different fields not covered here, or by improving on what we reported. We have only sketched here the well-documented field of neural networks, and ignored the still controversial area of protein folding, and for social effects the simulation of traffic jams (NAGEL and SCHRECKENBERG 1992).

Specific results open to checks by experts in the other fields are: the Azbel observation, of a characteristic age $X \simeq 103$ years for all humans independent of country and century, would mean that for a homogeneous population even under improved medical care the mean life span would not exceed X; the Gompertz ageing law has a microscopic explanation in the Penna model; sex is useful since it makes recessive genetic diseases less harmful for the individual; life after menopause may have genetic reasons and be connected with child care and reproduction risk at old age also for animals; lack of interleukin molecules might change an HIV infection into AIDS; heartbeat dynamics may be critical; deviations from Gaussian distributions for market fluctuations have a microscopic explanation; and a computer model for the alignment of countries in World War II and the Cold War was proposed.

Your comments might reach us fast through e-mail: suzana@if.uff.br; pmco @if.uff.br; and stauffer@thp.uni-koeln.de.

7 Appendices

7.1 Statistical Physics

For the non-physicists we summarize here the basic concepts of statistical physics and their Monte Carlo simulation.

Following Ludwig Boltzmann, statistical physicists assume that some quantum mechanical state of a system, having an energy E, is realized in thermal equilibrium with a probability proportional to $\exp(-\beta E)$. Here $\beta = 1/k_B T$ where k_B is Boltzmann's constant and T the absolute temperature. Thus if only two states exist with energy E_1 and E_2, the properly normalized probabilities are $p_1 = \exp(-\beta E_1)/(\exp(-\beta E_1) + \exp(-\beta E_2))$ with an analogous formula for the other level. For example, a magnetic dipole moment or spin in a magnetic field has the energy $-h$ if it points parallel to the field, and the energy $+h$ if it is oriented antiparallel to this field. Then the probabilities are $p = \exp(\pm\beta h)/(\exp(\beta h) + \exp(-\beta h))$ for pointing the spin in one of these two directions. We neglect here the possibility to orient the spin into any other direction, and also ignore other quantum mechanical effects; thus we should not call our energy a Hamiltonian.

In the Ising model, nearly as old as statistical physics, one assumes the spins on a lattice to interact as

$$E = -J \sum S_i S_j - h \sum_i S_i, \quad S_i = \pm 1$$

where the first sum goes over all pairs of nearest neighbours i and j on the lattice. The second term sums up the above mentioned energy of a spin in a magnetic field proportional to h, while in the first term J is usually called the exchange energy and gives the tendency of neighbouring spins to be parallel. We can interpret this Ising magnet also as a model for fluids, particularly near the liquid-vapour critical point, by regarding $S_i = +1$ as occupied and $S_i = -1$ as empty. (The liquid phase is the one with the majority of spins up, while the vapour has most spins down.) It then may be more plausible to use $\sigma_i = (1 + S_i)/2 = 0$ or 1 as a variable, which transforms the energy into

$$E = -4J \sum \sigma_i \sigma_j - \mu \sum_i \sigma_i + \text{Const}, \quad \mu = 2h + \text{const}$$

with the "chemical potential" μ and some lattice-dependent constants.

In Monte Carlo simulations we use random numbers r distributed homogeneously between zero and unity; if we want to do something with probability p, then we get a new random number r, and perform the desired operation only if $r < p$. General principles for Monte Carlo simulations are:

1: Select a new element, like a spin
2: Calculate the energy change ΔE involved e.g. with a spin flip
3: Perform the change with probability $\exp(-\beta\Delta E)$
4: Calculate the desired properties for your averages
5: Return to step 1

Repeating these five steps for all spins in the system constitutes one time unit (Monte Carlo step per spin). If instead of Ising spins some other model is simulated, then instead of a spin one may select another element (atom, ...), and instead of a spin flip one makes some other random change connected with this element. If in step 3 the energy change is negative, i.e. if the change lowers the energy, then the exponential function is larger than unity which means one always makes this change. This is the Metropolis method; the Glauber probability for making the change is $1/(1+\exp(\beta\Delta E))$ and gives the same equilibrium results but slightly different dynamics.

In an Ising model, we may calculate the local field $h_i = h+J\sum_j S_j$ acting on spin i, and point the spin parallel to h_i with probability $\exp(\beta h_i)/(\exp(\beta h_i)+\exp(-\beta h_i)) = 1/(1+\exp(-2\beta h_i))$. Since now $\Delta E = \pm 2h_i$ is the energy to flip the spin, this so-called "heat bath" method is on average equivalent to the Glauber method, as the reader may check by going through the various possibilities. The sum over j goes only over the neighbours interacting with i. In this way lattices with up to 7168^3 spins were simulated, and in two dimensions (with additional tricks, LINKE 1995) even up to one million times one million.

7.2 Programs for Biological Ageing

(listed after the references)

The two programs for the Penna model, sexual and asexual version, were written first in C and because of this bad youth do not follow the Fortran convention for integer names. Instead all variables must be declared explicitely at the beginning. The asexual version is a simplification of the sexual one and thus explained first. The senile author of this book is not familiar with C syntax; thus we present the Fortran versions.

The crucial array is called *dataf* and is basically the passport for the living females (no males yet for *datam*). In the rightmost (least significant) 6 bits of the 32-bit word *dataf* is stored the age, and in the next 6 bits to the left the number of mutations active at present.

Loop 7 goes over the time steps, from 1 to *maxstep*, with averages taken only over the last *medstep* iterations. The Verhulst factor is newly calculated for each new time step, but within one time step stays constant for all survivors.

After label 9 we read off from *dataf* (by shifts to the right and logical AND with the bit-string 111111) the age, active number of mutations, and remaining duration of pregnancy. Then the other crucial array, the genome *gen1f* for each individual, is checked for a possible new mutation, again by shifting *gen1f* to the right by age bits, followed by a logical AND with 1. The number *fnumber* of females at this age is decreased by one, and the age increased by one. A new random number *ibm* is drawn, and compared with the Verhulst factor. If the random number is smaller than the Verhulst factor (normalized to the interval from -2^{31} to $+2^{31}$ to speed up the program), or if the new number of mutations is no longer below the threshold *lim*, or if the age exceeds 32, then this individual dies; otherwise it survives and can give birth.

Death means that the information (*gen1f* and *dataf*) for the last individual simulated is placed into the array positions for the dead individual and the population is decreased by one, with special precautions if the last individual died.

Survival means that the number *fnumber* of females of the new age is increased by one, and the remaining duration of the pregnancy (if applicable) is decreased by one. The new pregnancy duration, number of active mutations, and age are stored in the passport *dataf*, via shifts to the left and logical OR operations. Now, if the past pregnancy is over, and the age is between the minimum and maximum ages for reproduction (*minage* and *fage*), births can happen.

At birth, each of the *nbirth* daughters produced in loop 12 gets the genome *gen1f* of the mother, with the exception that at one randomly selected bit position between 0 and 31 (obtained by looking at the five most significant bits of the random number *ibm*) the bit in the daughter genome *gene1* is set equal to one through a logical OR with a suitable array element *p*. The number of babies, *fnumber*(0), then is increased by one, and so is the population *fpop*. The new baby gets an empty passport *dataf*, while the mother gets the pregnancy restriction into her *dataf* word.

With the line of label 17 we increase our counter *i* by one and return to label 9 if this counter is smaller than the population. Loop 20, within the time loop but after all individuals have been dealt with at this time, collects the age distribution *fnumber* into the array *fynumber* during the latter part of the simulation.

For the sexual part the above statements are kept for the females (names ending with f) and added for the males, with variable names ending with m. But now breeding is more complicated (loop 10): Knowing that all males are equally disgusting though needed, the female makes 20 attempts to find randomly a suitable male. The selected male must have an age between the minimum and maximum male reproduction age (*minage* and *mage*), and should not have been selected at the same iteration by another female before (as marked by a breed bit in bit position 12 of the male passport *datam*).

The production of the child genome is also more complicated in the sexual case (loop 12). First at label 141 we determine randomly a number between 1 and 31 where the crossover for the female genome will occur; label 142 marks the same procedure for the male. Each individual has two genome words (*gen1f* and *gen2f* for the females, *gen1m* and *gen2m* for the males). The new genome words *g1* and *g2* are then the results of crossing: the first part of the 32 bits is taken from *gen1f*, the second part from *gen2f*, or the other way round (for the females, analogously for the males). Randomly, one of the two words *g1* and *g2* is selected for the egg *gene1* and the sperm *gene2*. Random mutations are added to both *gene1* and *gene2*.

Now the two gamete genomes *gene1* and *gene2* are fused to form the zygote: Randomly the offspring is male or female, the corresponding number of male or female babies is increased by one as well as the population, and *gene1* and *gene2* are stored as the new genome words. Three lines before label 17 the father is marked so as not to betray too soon the mother, while the mother is marked as pregnant like in the asexual case.

The crucial difference between dominant and recessive mutations, which makes sexual reproduction more reliable than asexual one, is dealt with earlier in the program. During initialization a dominance word is created giving d bits set among the 32 bits; mutations at these positions are dominant, all others recessive. In the loops over time and individuals, e.g. 2 and 3 lines after label 8, the additional number n of mutations no longer is obtained merely by reading the single passport word. Instead we now have two such words *gen1m* and *gen2m*, and so the number of bits set at the current age position is read from them and added up with the bit at that age position in the dominance word. If this sum n is equal to 2 or 3 it means that either both DNA strings *gen1m* and *gen2m* had the bit set, or one of them had it set and the mutation was dominant. This is the condition for a new genetic disease, increasing *nmut* by one. If the sum n is 1 or 0, the disease is recessive or non-existing, and *nmut* does not increase.

A minor point: bits 12 to 17 in the female passport word *dataf* indicate the remaining duration of the pregnancy until birth. During pregnancy no new pregnancy can be started, thus for each time interval the number stored in the six pregnancy bits is reduced by one, for each living pregnant female. (If we take the pregnancy duration as one time interval, this effect cancels out.)

In spite of these complications, each processor of a Cray-T3E updated about one individual per microsecond in the sexual case; the asexual program is nearly three times faster. Had we not used the passport words like dataf, we would have counted again and again at each age the number of set bits in the genome word, as in the original program (PENNA and STAUFFER 1995). This old method is simpler but slower.

7.3 Mutational Meltdown

If at every generation the survival probability of all individuals is reduced by a small fraction ϵ due to hereditary mutations, then it decays as $\exp(-\epsilon t)$ towards zero, and with a fixed number of offspring the species cannot possibly survive: *Mutational Meltdown* (LYNCH and GABRIEL 1990). On the other hand, if the offspring with probability 1/2 suffers no new mutations and has the same survival probability as the parent, while also with probability 1/2 its survival probability is reduced by 2ϵ, then these fluctuations allow the species to survive. However, if the population is small, even in the latter case where mutational meltdown in principle can be avoided, accidental fluctuations will eventually drive the population to extinction (PAL 1996).

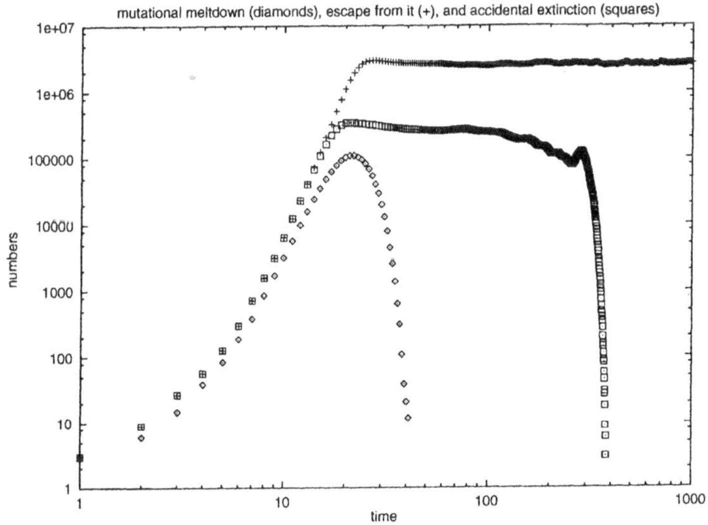

Fig.7.1 - Simple example of deterministic mutations leading to mutational meltdown (diamonds) and fluctuating mutations avoiding mutational meltdown (+); $maxpop = 10^7$ in both cases. With $maxpop = 10^6$ also for fluctuating mutations the population eventually vanished (squares).

Our third program and its results shown in fig.7.1 clarify these possibilities. Age structure is ignored, and instead of a bitstring each individual carries a real number *surviv* giving its inherited survival probability (initially = 1). In addition, Verhulst deaths proportional to the actual population size are taken into account by the factor *verhu*. Only if a random number *rand* is smaller than the product $surviv * verhu$ does the individual survive. In this case, its *surviv* is reduced by a factor $1 - eps = 0.95$, and it has $mchild = 3$ offspring inheriting this new survival rate. This case leads to mutational meltdown (diamonds in the figure).

If, on the other hand, we remove the c in front of the program line which with probability 1/2 sets *surviv* back to its non-reduced value, the species can survive if we take the carrying capacity *maxpop* as 10^7, even if we double *eps* to 0.1 in order to have the same average ϵ (plus signs in the figure.)

If now we reduce *maxpop* by a factor 10, the whole population decays in this case of fluctuating mutations (squares in the figure), but much slower than in the case of deterministic mutational meltdown.

These possibilities of mutational meltdown and how to avoid it are thus not restricted to ageing problems but rather general. They should also be discussed e.g. in Eigen's quasi-species model (EIGEN, McCASKILL and SCHUSTER 1990) which can be regarded as a bitstring model with hereditary mutations but without age structure, as attempted by MALARZ and TIGGEMANN 1998 and BAGNOLI and BEZZI 1998. This mutational meltdown is similar to the radioactive decay of a large number of atoms and is not a finite-size effect or fluctuation effect in the usual sense.

8 References

References ordered alphabetically, and according to year for multiple authors

AHMED E. (1996) J. Stat. Phys. **85**, 291.

ANDERSON P.W., ARROW K.J. and PINES D. (1988) eds., *The Economy as an Evolving Complex System*, Addison-Wesley, Reading MA.

ANDREEV A, BORODKIN L, and LEVANOVSKII M (1996) Historical Social Research **22**, 64.

ARGOLLO DE MENEZES M., RACCO A. and PENNA T.J.P. (1996) Physica A **233**, 221.

ARNEODO A., BACRY E., GRAVES P.V. and MUZY J.F. (1995) Phys. Rev. Lett. **74**, 3293.

ARNEODO A., DAUBENTOCARAFA Y., BACRY E. and GRAVES P.V. (1996) Physica D **96**, 291.

ARNEODO A., MUZY J.F. and SORNETTE D. (1998) Eur.J.Phys. B **2**, 277.

ARTHUR W.B. (1990) Scientific American **261**, February, 80.

ARTHUR, W. B., 1994. *Increasing Returns and Path Dependence in the Economy* (Ann Arbor, University of Michigan Press).

ARTHUR W.B., HOLLAND J.H., LE BARON B, PALMER R.G. and TAYLER P (1997) in: *The Economy as an Evolving Complex System II*, eds. W.B. Arthur, S. Durlauf, and D. Lane, Addison-Wesley, Redwood City CA, as summarized in CASTI J.L. (1997) *Would-Be Worlds*, Wiley, New York.

AUSLOOS M. (1998) Europhysics News **29**, 70.

AUSTAD S.N. (1994) Exp. Gerontol. **29**, 255.

AUSTAD S.N. (1997) Exp. Gerontol. **32**, 23.

AXELROD R and BENNET D.S. (1993) Brit. J. Polit. Sci. **23**, 211; see also (1998) **28**, 413 and R. AXELROD, *The Complexity of Cooperation: Agent-Based Models of Competition and Collaboration*, Princeton University Press, Princeton 1997.

AZBEL M.Ya. (1994) Proc. Natl. Acad. USA **91**, 12453.

AZBEL M.Ya. (1996) Proc. Roy. Soc. London B **263**, 1449.

AZBEL M.Ya. (1997) Phys.Repts. **288**, 545.

AZBEL M.Ya. (1998) Physica A **249**, 472 and Proc. Natl. Acad. Sci. USA **95**, 9037.

AZBEL M. Ya. (1999) preprint, Tel Aviv University

BACHELIER L. (1900) *Theorie de la Speculation*, Gauthier-Villars, Paris.

BAGNOLI F. and BEZZI M. (1998) Int. J. Mod. Phys. C. **9**, 999.

BAIER M., WERNER A., BANNERT N., METZNER K., and KURTH R (1995) Nature **378**, 563.

BAK P. (1997) *How Nature Works*, Oxford University Press, New York.

BAK P., TANG C and WIESENFELD K. (1987) Phys.Rev.Lett. **59**, 381.
BAK P. and CHEN K.(1991) Scientific American **264(1)**, 46 (January)
BAK P., PACZUSKI M. and SHUBIK M. (1997) Physica A **246**, 430.
BARETT, W.A., J. GAWEKE and SHELL K. (1989). *Economic complexity : chaos, sunspots, bubbles and nonlinearity*, Proceedings of the Fourth Internation Symposium in Economic Theory and Econometrics (Cambridge University Press, Cambridge).
BARKEMA G.T., HOWARD M.J., and CARDY J.L. (1996) Phys. Rev. E **92**, 2017.
BECKMAN K.B. and AMES B.N. (1998) Physiological Rev. **78**, 547.
BERNARDES A.T. (1995) J. Phys. I France **5**, 1501.
BERNARDES A.T. (1996a) *Annual Reviews of Computational Physics IV*, 359. Edited by D. Stauffer - World Scientific, Singapore.
BERNARDES A.T. (1996b) Ann. Physik **5**, 539.
BERNARDES A.T. (1997) J. Stat. Phys. **86**, 431.
BERNARDES A.T. and STAUFFER D. (1995) Int. Mod. Phys. C **6**, 789.
BERNARDES A.T. and ZORZENON DOS SANTOS, R.M. (1997) J. Theor. Biol. **186**, 173.
BERNARDES, A.T., MOREIRA, J.G., and CASTRO-E-SILVA, A. (1998) Europ. Physl J. B **1**, 393.
BERNSTEIN H. (1977) J. Theor. Biol. **69**, 371.
BERNTSEN K.N. (1996) Int. J. Mod. Phys. **C7**, 731.
BEZZI M., CELADA F., RUFFO S. and SEIDEN P.E. (1997) Physica A **245**, 145.
BODNAR A.G., OUELETTE M., HOLT, M.F.S.E, CHIU, C.P., MORIN, G.B., SHAY, J.W., LICHTSTEINER, S., and WRIGHT, W.E. (1998) Science **279**, 349.
BONABEAU E., THERAULAZ G., and DENEUBOURG J.L.(1995) Physica A **217**, 373.
BONABEAU E., THERAULAZ G., DENEUBOURG J.L., LIONI A., LIBERT F., SAUWENS C. and PASSERA L. (1998) Phys, Rev. E **57**, 5904.
BOUCHAUD, J.P. and POTTERS, M. (1997) *Theorie des Risques Financieres*, Alea-Saclay/Eyrolles.
BULDYREV S.V., AMARAL L.A.N., HAVLIN S., LESCHHORN H., MAASS P., SALINGER M.A., STANLEY, H.E., and STANLEY, M.H.R. (1997) J. Physique I **7**, 624.
CALDARELLI G., MARSILI M. and ZHANG Y.C. (1997) Europhys.Lett. **40**, 479.
CALLEN E. and SHAPERO D. (1974) Physics Today, July, 23.

CAREY J.R., LIEDO, P., OROZCO, D., and VAUPEL, J.W. (1992) Science **258**, 457.
CARLSON J.C. and RILEY J.C.M. (1998) Exp. Gerontol. **33**, 127.
CASTIGLIONE F, BERNASCHI, M. and SUCCI S. (1997) Int. J. Mod. Phys. C **8**, 527; see also SUCCI M., CASTIGLIONE F. and BERNASCHI M. (1997) Phys. Rev. Lett. **79**, 4493.
CEBRAT S. (1998) Physica A **258**, 493.
CEBRAT S. and KAKOL J. (1997) Int. J. Mod. Phys. C **8**, 417.
CEBRAT S. and DUDEK M.R. (1998) Europ. Phys. J. 3, 271; see also CEBRAT S., DUDEK M.R. and MACKIEWICZ P. (1998) Theory Bioscien. **117**, 78.
CELADA F. and SEIDEN P. (1992) Immunology Today **13**, No.2, 56 and J.Theor. Biol. **158**, 329.
CHAKRABARTI B.K. and MARJIT S. (1995) Indian J. Phys. **69** B, 681.
CHANG I. and STAUFFER D. (1998) Physica A, in press, Pusan National University.
CHARLESWORTH B. (1994) *Evolution in Age-Structured Populations*, 2nd edition, Cambridge University Press, Cambridge (UK).
CHARLESWORTH D., MORGAN M.T and CHARLESWORTH B. (1992/3) Genetic Res. (Cambridge), **59**, 49 (1992) and **61**, 39 (1993); J. Heredity **84**, 321 (1993).
CHATAGNY R. and CHOPARD B. (1997) International Conference on High Performance Computing and Networks, Vienna.
CHOWDHURY D. (1993) J. Theor. Biol. **165**, 135.
CHOWDHURY D. (1995) Indian J. Physics **69** B, 539.
CHOWDHURY D. and STAUFFER D. (1990) J. Stat. Phys. **59**, 1019.
CHOWDHURY D., SAHIMI M. and STAUFFER D. (1991) J. Theor. Biol. **152**, 263.
CHOWDHURY D., DESHPANDE V., and STAUFFER D. (1994) Int. J. Mod. Phys. C **5**, 1049.
CHOWDHURY D., SAHIMI M. and STAUFFER D. (1991) J. Theor. Biol. **152**, 263.
CHOWDHURY D. and STAUFFER D. (1999) Eur. Phys. J. B , in press, Cologne University
CLARK M.R. (1996) *Sex and the Origins of Death*, Oxford University Press, New York.
CONT R. and BOUCHAUD J.P. (1998) preprint; page 71 in BOUCHAUD and POTTERS 1997
COUTINHO A. (1989) Immunol. Rev. **110**, 63.
CREMER J. (1997) Physica A **246**, 377.

CRISTOFALO V.J. and CREPALDI G. (1998) Aging Clin. Exp. Res. **10**, 141 (conference proceedings).
CUEILLE S. and SIRE C. (1997) J. Phys. A **30**, L 791.
CURTSINGER, J.W., FUKUI, H.H, TOWNSEND, D.R. and VAUPEL, J.W. (1992) Science **258**, 461.
DASGUPTA S. (1994) J. Physique I **4**, 1563.
DASGUPTA S. (1997) Int. J. Mod. Phys. C **8**, 605.
DAWKINGS R. (1987) *The Blind Watchmaker*, Norton, New York
DAWKINGS R. (1995) Scientific American **273 (5)**, 62 (November).
DAYAN I., HAVLIN S. and STAUFFER D. (1988) J.Phys. A **21**, 2473.
DE ALMEIDA R.M.C., MOSS DE OLIVEIRA S. and PENNA T.J.P. (1998) Physica A **253**, 366.
DE ALMEIDA R.M.C. and MOUKARZEL C. (1998) Physica A **257**, 10.
DE BOER R.J., SEGEL L.A. and PERELSON A.S. (1992) J.Theor.Biol. **155**, 295.
DE BOER R.J., VAN DER LAAN, and HOGEWEG P. (1992) in *Thinking about Biology*, Varela, F.J. and Stein W., eds; Addison-Wesley, New York.
DE LIMA A.R., PENNA T.J.P. and DE OLIVEIRA P.M.C. (1997) Int. J. Mod. Phys. C **8**, 1073.
DE OLIVEIRA P.M.C. (1997) Computers in Physics **11**, 443.
DE OLIVEIRA P.M.C. and PENNA T.J.P. (1993) J. Stat. Phys. **73**, 789.
DE OLIVEIRA P.M.C. and PENNA T.J.P. (1994) Int. J. Mod. Phys. C **5**, 997.
DE OLIVEIRA P.M.C,, MOSS DE OLIVEIRA S. and STAUFFER D. (1997) Theory Bioscien. **116**, 3.
DE OLIVEIRA P.M.C., MOSS DE OLIVEIRA S., BERNARDES A.T. and STAUFFER D. (1999) Physica A **262**, 242.
DE OLIVEIRA P.M.C., MOSS DE OLIVEIRA S., BERNARDES A.T. and STAUFFER D. (1998) Lancet **352**, 911.
DE SOUSA A.O. and MOSS DE OLIVEIRA S. (1998) Eur. J. Phys. B , in press, and preprint, Universidade Federal Fluminense, Brazil.
DE SOUZA A.M.C. (1996) Sistemas Estatisticos Complexos e Mecanica Estatistica Nao Extensiva, doctoral thesis, CBPF, Rio de Janeiro.
DERRIDA B., BRAY A.J. and GODRECHE J. (1994) J. Phys. A **27**, L 357.
DREYER K. and HICKEY F.R. (1991) Am. J. Phys. **59**, 619.
DUPUIS H. (1997) Trends Tendances (Belgium) page 26 of 18 September 1997
EBELING W, SCHARNHORST A and KARMESHU, in Kertesz and Kondor (1998).
EIGEN, M., McCASKILL J. and SCHUSTER P (1990) Adv. Chem. Phys. **75**, 149.

EXCOFFIER L. (1997) La Recherche **302**, 82.

FAMA E.F. and FRENCH K.R. (1996) J. of Finance **51**, 55.

FARMER J.D., PACKARD N.H. and PERELSON A.S. (1986) Physica D **23**, 187.

FEHSENFELD K.M., SA MARTINS J.S., MOSS DE OLIVEIRA S. and BERNARDES A.T. (1998) Int. J. Mod. Phys. C **9**, 935.

FEINGOLD S.J. (1996) Physica **A231**, 499 and page 97 in: *Wolves of America Conference Proceedings*, November 1996 (Albany NY), Defenders of Wildlife, Washington, DC. See also (1998) Int. J. Mod. Phys. C 9, 295.

FISHER A., CALVET L., and MANDELBROT B. (1997), Cowles Foundation Discussion Paper 1166, Yale University, to be published in J. of Finance.

FOLLMER H (1974) J. Math. Economics **1**, 51.

FRACHEBOURG L, KRAPIVSKY P.L. and REDNER S. (1997) Phys.Rev. E **55**, 6684.

FRIES J F. (1980) N.Engl. J. Med. **303**, 130.

GALAM S. (1990) J. Stat. Phys. **61**, 943 and letter in Le Monde, May 30, 1997.

GALAM S. (1996) Physica A **230**, 174 and (1998) Brit. J. Polit. Sci. **28**, 411.

GALAM S. (1997) Physica A **238**, 66.

GALAM S., CHOPARD B., MASSELOT A. and DROZ M. (1998) Eur.Phys.J. B **4**, 529.

GENNOTTE, G., and LELAND H. (1990) Am. Economic Rev. **80**, 999.

GLANCE N.S. and HUBERMAN B.A. (1994) Scientific American **270**, March, 58.

GOPIKRISHNAN P., MEYER M., NUNES AMARAL L.A., and STANLEY H.E. (1998) Eur. Phys. J. B **3**, 139; see also T. LUX (1996) Appl. Financial Economics **6**, 463.

GOULD S.J. (1989) *Wonderful Life: The Burgess Shale and the Nature of History*, Norton, New York.

GRIMLEY EVANS J. et al. (1987) *Ageing: Science, medicine, and society*, conference proceedings, Philos. Trans. R. Soc. London (Biol.) 352, 1763-1920.

HARRIS T.E. (1989) *The Theory of Branching Processes*, Dover Publications Inc., New York.

HELLTHALER T. (1995) Int. J. Mod. Phys.C **6**, 845.

HEUMANN M. and HOTZEL M. (1995) J. Stat. Phys. **79**, 483.

HIRABAYASHI, T., TAKAYASU, H., MIURA H., and HAMADA. K (1993). Fractals **1**, 29.

HOFESTADT R. (1985) Masters Thesis, Bonner Informatik Berichte.

HOFFMANN J. and PEKALSKI A (1998) Int. J. Mod. Phys. C **9**, 633.

HOLBROOK N.J., MARTIN G.R. and LOCKSHIN R.A. (1996) *Cellular Ageing and Death*, Wiley-Liss, New York.
HORNOS J.E.M. and HORNOS Y.M.M. (1993) Phys. Rev. Lett. **71**, 4401.
HOPFIELD J.J. Proc. Natl. Acad. Sci. USA (1982) **79**, 2554.
ILINSKI K.N. and STEPANENKO A.S., preprint cond-mat/9806138
ISPOLATOV S., KRAPIVSKY P.L. and REDNER S. (1996) Phys. Rev. E **54**, 1274.
ISPOLATOV S., KRAPIVSKY P.L. and REDNER S. (1998) Eur. Phys. J. B **2**, 267.
ITO N. (1996a) Int. J. Mod. Phys. C **7**, 107.
ITO N. (1996b) Physica A **232**, 134.
JAN N. (1994) J. Stat. Phys. **77**, 915.
JERNE N.K. (1974) Ann. Immuno. **125** C, 373.
JOHANSEN A and SORNETTE D (1998) Eur.Phys.J. B **1**, 141 and in press.
JOHANSEN A , LEDOIT L and SORNETTE D (1999) preprint for Review of Financial Studies, Institute of Geophysics and Planetary Science, UCLA.
JOHNSON T.E. (1997) Exp. Gerontol. **32**, 11.
KAUFFMAN S.A. (1969) J. Theor. Biol. **22**, 437.
KAUFFMAN S.A. (1991) Scientific American **265 (2)**, 64 (August).
KAUFFMAN S.A (1993) *The Origins of Order*, and (1995) *At Home in the Universe*, Oxford University Press, New York.
KAUFMAN M, URBAIN and THOMAS (1985) J. Theor.Biol. **114**, 527.
KELLER L. and GENOUD M. (1997) Nature **389**, 958.
KERTESZ J. and KONDOR I. (1998) *Econophysics: An emerging science*, Kluwer, Dordrecht.
KIM G.R. and MARKOWITZ H.M. (1989) J. Portfolio Management **16**, Fall, 45.
KIZAKI S. and KATORI M. (1998) poster at "Percolation and Disordered Systems" Rauischholzhausen, July 1998.
KREBS J.R., ASHCROFT R. and WEBER M. (1978) Nature **271**, 539
KOHL R. (1997) Int. J. Mod. Phys. C **8**, 1309.
KOHRING G.A (1996) J. Physique I **6**, 301.
KOLTOVER V.K. (1997) J. Theor. Biol. **184**, 157; KOLTOVER, V.K., ANDRIANOVA Z.M.S. and IVANOVA A.N. (1993) Biol.Bull. Russian Acad.Sci. **20**, 95.
KOSCIELNY-BUNDE E., BUNDE A., HAVLIN S. and GOLDREICH Y. (1996) Physica A **231**, 393; KOSCIELNY-BUNDE E., BUNDE A., HAVLIN S., ROMAN H.E., GOLDREICH Y. and SCHELLNHUBER H.J. (1998) Phys. Rev. Lett. **81**, 729.
LEVY M, LEVY H. and SOLOMON S.(1994) Econ. Lett. **94**, 103.

LEVY M, LEVY H. and SOLOMON S.(1995) J. Physique I **5**, 1087.
LEVY M, SOLOMON S and RAM G. (1996) Int. J. Mod. Phys.C **7**, 65.
LEVY M, and SOLOMON S. (1996) Int. J. Mod. Phys.C **7**, 595.
LEVY M, and SOLOMON S. (1997) Physica A **242**, 90.
LI W. (1992) Int. J. Bifurcation and Chaos **2**, 137.
LI W. and KANEKO K. (1992) Europhys. Lett. **17**, 655.
LIN Y-J., SEROUDE L. and BENZER S. (1998) Science **282**, 943.
LINKE A., HEERMANN D.W., ALTEVOGT P., and SIEGERT M. (1995) Physica A **222**, 205.
LIPPERT K. and BEHN U. (1987) page 287 in *Annual Reviews of Computational Physics*, vol. V, edited by D. Stauffer, World Scientific, Singapore.
LUX T. (1997) J. Economic Dynamics and Control **22**, 1; (1998) J. Economic Behavior and Organization **33**, 143; LUX T. and MARCHESI, M., Nature, in press.
LYNCH M. and GABRIEL W. (1990) Evolution **44**, 1725.
LYRA M.L. (1998) in *Annual Reviews of Computational Physics*, vol. VI, edited by D. Stauffer, World Scientific, Singapore.
MACLEOD S. and JAN N. (1998) Int. J. Mod. Phys. C **9**, 289.
MADDOX J. (1994) Nature **367**, 111.
MALARZ K. and TIGGEMANN D. (1998) Int. J. Mod. Phys. C **9**, 481.
MANDELBROT B. (1966/67) J.Business Univ. Chicago **39**, 242 and **40**, 393.
MANDELBROT B.B. (1982) *The Fractal Geometry of Nature*, Freeman, San Francisco
MANDELBROT B.B. (1997) *Fractals and scaling in finance: discontinuity, concentration, risk*, Springer, New York
MANNA S.S., HERRMANN H.J. and LANDAU D.P. (1992) J. Stat. Phys. **66**, 1155.
MANTEGNA R.N. and STANLEY H.E. (1997) Physica A **239**, 255.
MAKOWIEC D. (1997) Physica A **245**, 99.
MARSH H. and KASUYA (1984) *Report of the International Whaling Commission*, (Special Issue 6) - page 163 of WACHTER and FINCH (1997).
MARSILI M. and ZHANG Y.C. (1997) Physica A **245**, 181.
MARSILI M., MASLOV S. and ZHANG Y.C. (1998) Physica A **253**, 403.
MARTIEN P., POPE S.C., SCOTT P.L. and SHAW R. (1995) Phys. Lett. **110** A, 339.
MARTINS S.G.F. and PENNA T.J.P. (1998) Int. J. Mod. Phys. C **9**, 491.
MARTINS S.G.F., PENNA T.J.P. and DE OLIVEIRA P.M.C. (1999) Eur. Phys. J. B, in press, Universidade Federal Fluminense, Brazil.
MASORO E.J. and BERGAMINI E.(1997) (eds.) Aging Clin.Exp.Res. **9**, 428-446.

McCLEARN G.E.(1997) Exp. Gerontol. **32**, 3.
MEISGEN F. (1997) Int. J. Mod. Phys. C **8**, 575.
MEYER H. (1995) Int. J. Mod. Phys. C **6**, 765.
MOSS DE OLIVEIRA (1998) Physica A **257**, 465.
MOSS DE OLIVEIRA S., PENNA T.J.P. and STAUFFER D. (1995) Physica A **215**, 298.
MOSS DE OLIVEIRA S., DE OLIVEIRA P.M.C. and STAUFFER D. (1995) Physica A **221**, 453.
MOSS DE OLIVEIRA S., DE OLIVEIRA P.M.C. and STAUFFER D. (1996) Braz.J.Phys. **26**, 626.
MOSS DE OLIVEIRA S.M., MEDEIROS G.A., DE OLIVEIRA P.M.C. and STAUFFER D. (1998) Int. J. Mod. Phys. C **9**, 809.
MOSS DE OLIVEIRA S., BERNARDES A.T. and SA MARTINS J.S. (1999) Eur. Phys. J. B **7** in press, Universidade Federal Fluminense, Brazil.
MUELLER L.D., NUSBAUM T.J. and ROSE M.R. (1995) Exp. Gerontol. **30**, 553.
MUELLER L.D. and ROSE M.R. (1996) Proc.Natl.Acad.Sci. USA **93**, 15249.
MULLER-HILL B. (1993) Nature **362**, 491.
NAGEL K. and SCHRECKENBERG M. (1992) J. Physique I **2**, 2221.
NEUMANN A.U. (1989) Physica A **162**, 1.
NEUMANN A.U. and WEISBUCH G. (1992) Bull.Math.Biol. **54**, 21.
NOWAK A., SZAMREIJ J. and LATANE B. (1990) Psychol. Rev. **97**, 362 (1990).
ORLEAN A. (1991) La Recherche **22**, 668.
OZAWA T. (1997) Physiological Review **77**, 425.
PACKER C., TARTAR M. and COLLINS A. (1998) Nature **392**, 807.
PAL K.F. (1996) Int. J. Mod. Phys. C **7**, 899.
PALMER, R.G., ARTHUR W.B., HOLLAND J.H., LEBARON B., and TAYLER P. (1994) Physica D **75**, 264.
PALMER R.J. and ADLER J. (1998) poster P004/72 at 20th IUPAP Int. Conf. Statistical Physics, Paris, July 1998, and priv. comm.
PAMILO P., NEI M. and LI W.H. (1987) Genet. Res., Camb. **49**, 135.
PANDEY R.B. (1996) manuscript for conference "Scientific Computing in Europe", SCE 96, September 2-4, 1996, Dublin, Ireland, edited by H.J. Ruskin, R. O'Connor, and Y. Feng.
PANDEY R.B. (1998) Theory Bioscien. **117**, 32.
PANDEY R.B. and STAUFFER D. (1989) J. Physique **50**, 1.
PANDEY R.B. and STAUFFER D. (1990) J. Stat. Phys. **61**, 235.
PAPA A.R.R. and TSALLIS C. (1995) Physica A **233**, 85.
PAPA A.R.R. and TSALLIS C. (1998) Phys.Rev. E **57**, 3923.

PARETO V. (1897) *Cours d'Economique Politique*, vol. 2. reprinted in: *Oevres Completes de Vilfredo Pareto*, I. edited by Bousquet G.H. and Busino, G. (1964), Libraire Droz, Geneve.
PARTRIDGE L. and BARTON N.H. (1993) Nature **362**, 305.
PENG C.-K., BULDYREV S.V., GOLDBERGER A.L., HAVLIN S., SCIORTINO F., SIMONS M. and STANLEY H.E. (1992) Nature **356**, 168.
PENG C.-K., MIETUS J., HAUSSDORFF J.M., HAVLIN S., STANLEY H.E. and GOLDBERGER A.L. (1993) Phys. Rev. Lett. **70**, 1343.
PENNA T.J.P. (1995) J.Stat.Phys., **78**, 1629.
PENNA T.J.P. and STAUFFER D. (1995) Int. J. Mod. Phys. C **6**, 233.
PENNA T.J.P. and MOSS DE OLIVEIRA S. (1995) J. Physique I **5**, 1697.
PENNA T.J.P., DE OLIVEIRA P.M.C., SARTORELLI J.C., GONÇALVES W.M. and PINTO R.D. (1995) Phys. Rev. E **52**, R2168.
PENNA T.J.P. and STAUFFER D. (1996) Zeits. Phys.B **101**, 469.
PENNA T.J.P. and WOLF D. (1997) Theory Bioscien. **116**, 118.
PENNA T.J.P., MOSS DE OLIVEIRA S. and STAUFFER D. (1995) Phys. Rev. E **R52**, 3309.
PENNA T. J. P., RACCO A. and ARGOLLO DE MENEZES M. (1998) Comput. Phys. Comm., in press (CCP 1998 Granada) Boston University.
PERAZZO, R.P.J., REICH S.L., SCHWARZER J., and VIRASORO M.A. (1995). Chaos, Solitons and Fractals **6**, 455.
PERELSON A.S. and OSTER G.F. (1979) J. Theor. Biol. **81**, 645.
PERELSON A.S. and WEISBUCH G. (1997) Rev. Mod. Phys. **69**, 1219.
PEREZ-CAMPO R., MOPEZ-TORRES M., CADENAS S., ROJAS C., BARJA G. (1998) J. Comp. Physiol. B **168**, 149.
PERLS T.T. (1995) Scientific American **272**, January, 50; see also (1997) Med. Hypotheses **49**, 405.
PINTO R.D., GONÇALVES W.M., SARTORELLI J.C. and DE OLIVEIRA M.J. Phys. Rev. E **52**, 6892 (1995).
PLETCHER S.D. and CURTSINGER J.W. (1998) Evolution **52**, 454.
PUHL H., STAUFFER D. and ROUX S. (1995) Physica A **221**, 445.
RAAB A. (1998) J. Stat. Phys. **91**, 1055.
RACCO A., ARGOLLO DE MENEZES M. and PENNA T.J.P. (1998) Theory Bioscien. **117**, 101.
RAY T.S. (1994) J. Stat. Phys. **74**, 929.
REDFIELD R.J. (1994) Nature **369**, 145.
RICH S.M., LICHT M.C., HUDSON, R.R., and AYALA, F.J. (1998) Proc. Natl. Acad.Sci. USA **95**, 4425.
ROEHNER B. (1989). Environment and Planning A **21**, 289.
ROSE G.A. (1993) Nature **366**, 458.

ROSE M. (1991) *Evolutionary Biology of Aging*, Oxford University Press, New York.
ROUGHGARDEN J. (1996) *Theory of Population Genetics and Evolutionary Ecology*, 2nd edition by Prentice-Hall.
RUBIN H. (1997) Mech. Ageing Dev. **98**, 1.
RUCKAMP R (1998) Staatsexamensarbeit, Cologne University
SAHIMI M. and STAUFFER D. (1993) Phys. Rev. Lett. **71**, 4271.
SA MARTINS J.S. and MOSS DE OLIVEIRA S. (1998) Int. J. Mod. Phys. C **9**, 421.
SARTORELLI J.C., W.M. GONÇALVES and R.D. PINTO (1994) Phys. Rev. E **49**, 3963.
SATO A.H. and TAKAYASU H. (1998) Physica A **250**, 231.
SCHNEIDER J., CEBRAT S. and STAUFFER D. (1998) Int. J. Mod. Phys. C **9**, 721; see also SMITH D.W.E, (1989) Biol. Rev. **64**, 1.
SCHWEITZER F. (1997) (ed.) *Self-Organization of Complex Structures: From Individual to Collective Dynamics*, Gordon and Breach, Amsterdam.
SCIENCE (1998) *The Evolution of Sex*, multi-authored special section in Science **281**, 1979 - 2008.
SEDIVY J.M. (1998) Proc. Natl. Acad. Sci. USA **95**, 9078. SHAW R. (1984) *The Dripping Faucet as a Model Chaotic System*, Aerial Press, Santa Cruz, CA.
SHILLER, R.J., (1989). *Market volatility*, The MIT Press, Cambridge, MA.
SIMON H. (1957) *Models of Man*, Wiley, New York.
SORNETTE D, (1998) Physics Reports, **297**, 239.
SORNETTE D. and CONT R. (1997) J. Physique I **7**, 431.
SORNETTE D, JOHANSEN A and BOUCHAUD J.P. (1996) J.Physique I **6**, 167.
SORNETTE D and JOHANSEN A (1997) Physica A **245**, 411 and (1998) Physica A **261**, 581.
STANLEY H.E., BULDYREV S.V., GOLDBERGER A.L., GOLDBERGER Z.D., HAVLIN S., MATEGNA R.N., OSSADNIK S.M., PENG C.-K. and SIMONS M. (1994) Physica A **205**, 214.
STAUFFER D. (1994) Braz. J. Phys. **24**, 900.
STAUFFER D. (1996) Comput. in Phys. **10**, 341 and (1999) Comput.Sci.Eng.
STAUFFER D. and WEISBUCH G. (1992) Physica A **180**, 42.
STAUFFER D. and JAN, N. (1993) unpublished, rejected by Nature and by Evolution.
STAUFFER D. and SAHIMI M. (1993) Int. J. Mod. Phys. C **4**, 401.
STAUFFER D., DE OLIVEIRA P.M.C., MOSS DE OLIVEIRA S.M. and ZORZENON DOS SANTOS R.M. (1996) Physica A **231**, 504.

STAUFFER D. and SORNETTE D. (1998) Physica A **252**, 271.
STAUFFER D. and PENNA T.J.P.P. (1998) Physica A **256**, 284.
STAUFFER D., DE OLIVEIRA P.M.C. and BERNARDES A.T. (1999) preprint for Int. J. Theor. Appl. Finance, Cologne University.
STEARNS, S.C. (1987) ed., *The Evolution of Sex and its Consequences*, Birkhauser, Basel.
STEIGLITZ K., HONIG M.L. and COHEN L.M. (1996) chapter 1 in *Market-Based Control: A Paradigm for Distributed Resource Allocation.* Edited by S. Clearwater, World Scientific, Hong Kong; STEIGLITZ K. and SHAPIRO D. (1996) "Simulating the Madness of Crowds: Price Bubbles in an Auctian-Mediated Robot Market", fall meeting of the Economic Sciemnce Association, Tucson, Arizona October 1996; STEIGLITZ K. and O'CALLAGHAN L.I., "Microsimulation of Markets and Endogenous Price Bubbles", Third Interntl. Conf. on Computing in Economics and Finance, Stanford 1997.
STREHLER S.L. and MILDVAN A.S. (1960) Science **132**, 14.
SUTTON, P., GEORGALLAS, A., HUNTER, D.L., JAN, N., NASH R.J., and QUINN, J.P. (1998) Int. J. Mod. Phys. C **9**, 547.
SUZUKI J. and KANEKO K. (1994) Physica D **75**, 328.
SZILARD L. (1959) Proc. Natl. Acad. Sci. USA **45**, 30.
TANG L.H. and TIAN G.S. (1998) Hong Kong Baptist University, preprint cond-mat/9811114.
TAKAYASU, H., H. MIURA, HIRABASHI T. and HAMADA K. (1992) Physica A **184**, 127.
THATCHER A.R., KANNISTO V. and VAUPEL J.W. (1998) *The Force of Mortality and Ages 80 to 120*, Odense University Press, Odense.
THOMS J., DONAHUE P. and JAN N. (1995) J. Physique I **5**, 935.
THOMS J., DONAHUE P., HUNTER D. and JAN N. (1995) J. Physique I **5**, 1689.
TOLEDO PIZA A.F.R. (1996) Physica A **242**, 195.
TOME T. and DRUGOWICH DE FELICIO J.B. (1996) Phys. Rev. E **53**, 3976.
TOULOUSE G. and BOK J. (1978). Revue Francaise de Sociologie **19**, 391.
TSALLIS C., (1997) Physics World **10** (No.7) 42.
TSALLIS C., PLASTINO A.R. and ZHENG W.-M. (1997) Chaos, Solitons and Fractals **8**, 885 (1997).
TSALLIS C. DE SOUZA A.M.C. and CURADO E.M.F. (1995) Chaos, Solitons and Fractals **6**, 561.
VANDEWALLE N. (1997) Physica A **245**, 113.
VANDEWALLE N. and AUSLOOS M. (1995) page 45 in *Annual Reviews of Computational Physics*, vol. III, D. Stauffer, ed., World Scientific, Singapore

1995.
VANDEWALLE N. and AUSLOOS M. (1997) Physica A **246**, 454 and (1998) Eur. Phys. J. B **4**, 257.
VANDEWALLE N., BOVEROUX P., MINGUET A. and AUSLOOS M. (1997) Physica A **255**, 201 and (1998) Eur.Phys.J B **4**, 139
VARIAN, H.R. (1989). Difference of opinion in financial markets, in: *Financial Risk : Theory, Evidence and Implications*, Proceedings of the Eleventh Annual Economic Policy Conference of the Federal Reserve Bank of St. Louis, edited by Courtenay C. Stone (Kluwer Academic Publishers, Boston).
VAUPEL J.W. et al (1998) Science **280**, 855..
VOLLMAR S. and DASGUPTA, S. (1994) J. Physique I **4**, 817.
VOSS R.F. (1992) Phys. Rev. Lett. **68**, 3805.
WACHTER K.W. and FINCH C.E. (1997) *Between Zeus and the Salmon. The Biodemography of Longevity*, National Academy Press, Washington DC.
WALLACE D.C. (1994) Proc. Natl. Acad. Sci. USA **91**, 8739.
WALLACE D.C. (1997) Scientific American **277** (August) 22.
WEALE R.A. (1997) Mech. Ageing Dev. **97**, 55.
WEISBUCH G. and ATLAN H. (1988) J.Phys. A **21**, L 189.
WEISBUCH G., STANLEY E.A., DUCHATEAU-NGUYEN G., ANTONA, M and CLEMENT-PITIOT, H. (1997) Theory Bioscien. **116**, 97.
WEISBUCH G., KIRMAN A. and HERREINER D. (1998) preprint for J. Theor. Economy.
WILLIAMS G.C. (1957) Evolution **11**, 398.
WOLFRAM S (1986) *Theory and Applications of Cellular Automata*, World Scientific, Singapore.
WU X. and SCHELLY A. Physica D **40**, 433 (1989).
YEPES H.N.N., BRITO A.L.S., VARGAS C.A. and VICENTE L.A. (1989) Eur. J. Phys. **10**, 99.
ZEBENDE G.F., PENNA T.J.P. and DE OLIVEIRA, P.M.C. (1998). Phys. Rev. E **57**, 3311 and Physica A **257**, 136.
ZHANG W.B. (1991) *Synergetic Economics.* Springer, Berlin-Heidelberg.
ZHANG Y.C., SERVA M. and POLIKARPOV M. (1990) J Stat. Phys. **58**, 849.
ZORZENON DOS SANTOS R.M. (1998) in *Annual Reviews of Computational Physics*, vol. VI, edited by D. Stauffer, World Scientific, Singapore 1998.
ZORZENON DOS SANTOS, R.M. and BERNARDES, A.T. (1995) Physica A **219**, 1.
ZORZENON DOS SANTOS, R.M. and BERNARDES, A.T. (1998) Phys. Rev. Letters **81**, 3034.

9 Programs

9.1 Asexual Penna ageing model

```
c     asexual version (females only)
      implicit none
      integer popdim
      parameter(popdim=5000  )
      integer popmax,inipop,maxstep,medstep,minage,fage,lim,fmut,birth,
     1 n6,t,verhu,i,p,seed,ibm,gene1,fa,n,imut,age,nmut,
     2 fpop, fymed(0:32),fnumber(0:32),bit(0:32),
     3 gen1f(popdim),dataf(popdim)
      parameter(popmax=2000    , inipop=200   , maxstep=10000,
     1 medstep=5000, minage= 8, fage=32, lim=3,
     2 fmut=1, birth=3, seed=1, n6=63)
c
      ibm=2*seed-1
      print *, popmax,inipop,maxstep,medstep,minage,
     1         fage,lim,fmut,birth,seed
      fpop=inipop
      bit(0)=1
      do 2 i=0,32
        if(i.gt.0) bit(i)=ishft(bit(i-1),1)
        fymed(i)=0
        fnumber(i)=0
 2      ibm=ibm*16807
      fnumber(0)=fpop
      do 6 i=1,fpop
        dataf(i)=0
 6      gen1f(i)=0
c     dataf: age at bits 0 to 5, mutations at 6 to 11
c
      do 7 t=1,maxstep
      verhu=(fpop*2.0/popmax-1.0)*2147483647
      if(t.eq.maxstep) print *, t,fpop
      i=1
      fa=fpop
 9    age =iand(n6,dataf(i))
      nmut=iand(n6,ishft(dataf(i),-6))
      n=iand(1,ishft(gen1f(i),-age))
      if(n.ge.1) nmut=nmut+1
      fnumber(age)=fnumber(age)-1
      age=age+1
      ibm=ibm*16807
      if(ibm.lt.verhu.or.nmut.ge.lim.or.age.eq.33) then
c     death
        if(fpop.le.1) goto 1
        gen1f(i)=gen1f(fpop)
        dataf(i)=dataf(fpop)
        fpop=fpop-1
        if(fpop.ge.fa) then
          i=i+1
        else
          fa=fa-1
        endif
      else
```

```
c     survival
        fnumber(age)=fnumber(age)+1
        dataf(i)=ior(age,ishft(nmut,6))
        if(age.ge.minage.and.age.le.fage) then
 11       do 12 n=1,birth
c         birth
            gene1=gen1f(i)
            do 13 imut=1,fmut
              ibm=ibm*16807
              p=bit(ishft(ibm,-27))
c             mutations in mother
 13           gene1=ior(gene1,p)
              fnumber(0)=fnumber(0)+1
              fpop=fpop+1
              if(fpop.gt.popdim) goto 1
              dataf(fpop)=0
              gen1f(fpop)=gene1
 12       continue
c         if(female suitable) then
        endif
        i=i+1
      endif
c     if(death) .. else (survival, birth) ..
      if(i.le.fa) goto 9
c     end of selection and birth, now start averages
      if(t.lt.maxstep-medstep) goto 7
      do 10 i=0,32
 10     fymed(i)=fymed(i)+fnumber(i)
 7    continue
      print 100, (i,fymed(i)*1.0/medstep,i=0,32)
 100  format(1x,i3,f11.2)
      stop
 1    print *, 'error'
      stop
      end
```

9.2 Sexual Penna ageing model

```
c     sexual version with males and females different
      implicit none
      integer popdim
      parameter(popdim=4000 )
      integer popmax,inipop,maxstep,medstep,minage,mage,fage,
     1 lim,mmut,fmut,birth,pregn,d,m12,n6,t,verhu,dominance,mstep,
     2 i,p,seed,ibm, ii, mc, nc, g1, g2, gene1, gene2, fa,
     3 n,agei,cr,imut,q12, age,nmut,breed,preg,pop,mpop,fpop,
     4 mnumber(0:32),fnumber(0:32),bit(0:32),mymed(0:32),fymed(0:32),
     5 gen1m(popdim),gen2m(popdim),gen1f(popdim),gen2f(popdim),
     6 datam(popdim),dataf(popdim)
      real factor,av
      parameter(popmax=1000   , inipop=100   , maxstep= 20480,
     1 medstep=10240, minage=8, mage=32, fage=32, lim=3, mmut=1,
     2 fmut=1, birth=4, pregn=1, d= 6, seed=1)
      data q12/-4097/, m12/4096/, n6/63/
c     only bad mutations, only crossover. Penna model with sex and pregnancy
c     Parameters: Verhulst pop., initial pop, array dimension, maximum time,
c     averages, reprod. ages: min, max male, max female, threshold,male mut,
c     femalemut, birth rate, waiting period, dominant genes
c
      ibm=2*seed-1
      factor=1.0/2147483647
      print *, popmax,inipop,maxstep,medstep,minage,mage,fage,lim,mmut,
     1          fmut,birth,pregn,d,seed
      mstep=maxstep-medstep
      av=1.0/medstep
      mpop=inipop
      fpop=inipop
      bit(0)=1
      do 2 i=0,32
        if(i.gt.0) bit(i)=ishft(bit(i-1),1)
        mymed(i)=0
        fymed(i)=0
        mnumber(i)=0
        fnumber(i)=0
 2      ibm=ibm*16807
      dominance=0
      do 4 i=1,d
 5      ibm=ibm*16807
        p=bit(ishft(ibm,-27))
        if(iand(dominance,p).ne.0) goto 5
 4      dominance=ior(dominance,p)
c     d randomly set bits in DOMINANCE determine position of dominant mutations
      mnumber(0)=mpop
      fnumber(0)=mpop
      do 6 i=1,mpop
        datam(i)=0
        dataf(i)=0
        gen1m(i)=0
        gen1f(i)=0
        gen2m(i)=0
 6      gen2f(i)=0
c     datam and dataf: age at bits 0 to 5, mutations at 6 to 11, pregnancy
c     at 12 to 17, male exhaustion at bit 12
c
      do 7 t=1,maxstep
      pop=fpop+mpop
      verhu=(pop*2.0/popmax-1.0)*2147483647
      if(t.eq.maxstep) print *, t,mpop,fpop,pop
      i=1
 8    age=iand(n6,datam(i))
      nmut=iand(n6,ishft(datam(i),-6))
      n=iand(1,ishft(gen1m(i),-age))+iand(1,ishft(gen2m(i),-age))+
     1  iand(1,ishft(dominance,-age))
```

```
      if(n.ge.2) nmut=nmut+1
      mnumber(age)=mnumber(age)-1
      age=age+1
      ibm=ibm*16807
      if(ibm.lt.verhu.or.nmut.ge.lim.or.age.eq.33) then
c     death
        if(mpop.le.1) goto 1
        gen1m(i)=gen1m(mpop)
        gen2m(i)=gen2m(mpop)
        datam(i)=datam(mpop)
        mpop=mpop-1
      else
c     survival
        mnumber(age)=mnumber(age)+1
        datam(i)=ior(age,ishft(nmut,6))
        i=i+1
      endif
      if(i.le.mpop) goto 8
      i=1
      fa=fpop
 9    age =iand(n6,dataf(i))
      nmut=iand(n6,ishft(dataf(i),-6))
      preg=iand(n6,ishft(dataf(i),-12))
      n=iand(1,ishft(gen1f(i),-age))+iand(1,ishft(gen2f(i),-age))+
     1  iand(1,ishft(dominance,-age))
      if(n.ge.2) nmut=nmut+1
      fnumber(age)=fnumber(age)-1
      age=age+1
      ibm=ibm*16807
      if(ibm.lt.verhu.or.nmut.ge.lim.or.age.eq.33) then
c     death
        if(fpop.le.1) goto 1
        gen1f(i)=gen1f(fpop)
        gen2f(i)=gen2f(fpop)
        dataf(i)=dataf(fpop)
        fpop=fpop-1
        if(fpop.ge.fa) then
          i=i+1
        else
          fa=fa-1
        endif
      else
c     survival
        fnumber(age)=fnumber(age)+1
        if(preg.gt.0) preg=preg-1
        dataf(i)=ior(ishft(preg,12),ior(age,ishft(nmut,6)))
        if(age.ge.minage.and.age.le.fage.and.preg.eq.0) then
c     breeding
          do 10 n=1,20
            ibm=ibm*16807
            if(ibm.lt.0) ibm=ibm+2147483647+1
            ii=ibm*factor*mpop+1
            if(ii.gt.mpop.or.ii.le.0) goto 10
            agei=iand(n6,datam(ii))
            breed=iand(1,ishft(datam(ii),-12))
            if(agei.gt.mage.or.agei.lt.minage.or.breed.ne.0) goto 10
            goto 11
c           partner found
 10       continue
c         no partner found
          goto 17
 11       do 12 n=1,birth
c         birth
 141        ibm=ibm*65549
            cr=ishft(ibm,-27)
```

```
             if(cr.le.0) goto 141
c            female crossing
             mc=bit(cr)-1
             nc=not(mc)
             g1=ior(iand(gen1f(i),mc),iand(gen2f(i),nc))
             g2=ior(iand(gen1f(i),nc),iand(gen2f(i),mc))
             ibm=ibm*65539
             gene1=g1
             if(ibm.lt.0) gene1=g2
             do 13 imut=1,fmut
               ibm=ibm*16807
               p=bit(ishft(ibm,-27))
c              mutations in mother
 13            gene1=ior(gene1,p)
 142         ibm=ibm*65539
             cr=ishft(ibm,-27)
             if(cr.le.0) goto 142
c            male crossing
             mc=bit(cr)-1
             nc=not(mc)
             g1=ior(iand(gen1m(ii),mc),iand(gen2m(ii),nc))
             g2=ior(iand(gen1m(ii),nc),iand(gen2m(ii),mc))
             ibm=ibm*65539
             gene2=g1
             if(ibm.lt.0) gene2=g2
             do 16 imut=1,mmut
               ibm=ibm*16807
               p=bit(ishft(ibm,-27))
c              mutations in father
 16            gene2=ior(gene2,p)
             ibm=ibm*16807
c            now fusion of gene1 and gene2 into child
             if(ibm.lt.0) then
               fnumber(0)=fnumber(0)+1
               fpop=fpop+1
               if(fpop.gt.popdim) goto 1
               dataf(fpop)=0
               gen1f(fpop)=gene1
               gen2f(fpop)=gene2
             else
               mnumber(0)=mnumber(0)+1
               mpop=mpop+1
               if(mpop.gt.popdim) goto 1
               datam(mpop)=0
               gen1m(mpop)=gene1
               gen2m(mpop)=gene2
             endif
 12        continue
           dataf(i )=ior(dataf(i),ishft(pregn,12))
           datam(ii)=ior(datam(ii),m12)
c          if(female suitable) then
         endif
 17      i=i+1
       endif
c      if(death) .. else (survival, birth) ..
       if(i.le.fa) goto 9
c      end of selection and birth, now clear data for next year
       do 15 i=1,mpop
c      males no longer breed
 15    datam(i)=iand(datam(i),q12)
       if(t.lt.mstep) goto 7
c      start averages
       do 20 i=0,32
         mymed(i)=mymed(i)+mnumber(i)
 20      fymed(i)=fymed(i)+fnumber(i)
```

```
7     continue
      print 100, (i,mnumber(i),fnumber(i),av*mymed(i),av*fymed(i),
     1  i=0,32)
100   format(1x,i3,2i9,2f9.2)
      stop
1     print *, 'error'
      stop
      end
```

9.3 Mutational meltdown

```
c     mutation meltdown exercise
      parameter(npop0=1,maxpop=1000000*npop0)
      dimension surviv(maxpop)
      data maxtime/1000/,mchild/3/,iseed/123/,eps/0.10/
c     input parameters: maxtime iterations, birthrate mchild,
c     random number seed iseed, reduction of yearly survival probability eps
      print *,npop0,maxtime,mchild,iseed,eps
      factor=1.0/maxpop
      asp=rand(2*iseed-1)
c     initialize here your random number generator, called rand(0) now
      npop = npop0
      do i = 1,npop
         surviv(i) = 1.0
      end do
      do itime = 1,maxtime
         verhu=1.0-npop*factor
         asp = 0.0
         icount = 0
         do i = 1,npop
            s = surviv(i)
            asp = asp + s
            if (rand(0) .le. s*verhu) then
             icount = icount + 1
             surviv(icount) = s*(1.0 - eps)
             if (rand(0) .gt. 0.5) surviv(icount) = s
            endif
         end do
c        compute average survival probability asp
         asp = asp/npop
         npop = icount
c        random survival finished; now deterministic birth
         print *, itime,mchild*npop,asp
         if (npop .le. 0) stop
         do i = npop,1,-1
            do ichild = mchild,1,-1
               surviv(i*mchild+ichild-mchild) = surviv(i)
            end do
         end do
         npop = mchild*npop
      end do
      stop
      end
```

Index

GPSR Compliance

The European Union's (EU) General Product Safety Regulation (GPSR) is a set of rules that requires consumer products to be safe and our obligations to ensure this.

If you have any concerns about our products, you can contact us on ProductSafety@springernature.com

In case Publisher is established outside the EU, the EU authorized representative is:

Springer Nature Customer Service Center GmbH
Europaplatz 3
69115 Heidelberg, Germany

Batch number: 10726013

Printed by Printforce, the Netherlands